Innovations in Agricultural Engineering and Mechanization for Sustainable Agriculture, Forestry and Food Production

Innovations in Agricultural Engineering and Mechanization for Sustainable Agriculture, Forestry and Food Production

Editors

Marcello Biocca
Roberto Fanigliulo

MDPI

Basel • Beijing • Wuhan • Barcelona • Belgrade • Novi Sad • Cluj • Manchester

Editors
Marcello Biocca
Centro di ricerca Ingegneria e
Trasformazioni Agroalimentari
CREA
Monterotondo
Italy

Roberto Fanigliulo
Centro di ricerca Ingegneria e
Trasformazioni Agroalimentari
CREA
Monterotondo
Italy

Editorial Office
MDPI
St. Alban-Anlage 66
4052 Basel, Switzerland

This is a reprint of articles from the Special Issue published online in the open access journal *AgriEngineering* (ISSN 2624-7402) (available at: www.mdpi.com/journal/agriengineering/special_issues/InnovationsAgriculturalEngineering).

For citation purposes, cite each article independently as indicated on the article page online and as indicated below:

Lastname, A.A.; Lastname, B.B. Article Title. *Journal Name* **Year**, *Volume Number*, Page Range.

ISBN 978-3-7258-0548-8 (Hbk)
ISBN 978-3-7258-0547-1 (PDF)
doi.org/10.3390/books978-3-7258-0547-1

Contents

About the Editors vii

Preface ix

Thiago Orlando Costa Barboza, Marcelo Araújo Junqueira Ferraz, Cristiane Pilon, George Vellidis, Taynara Tuany Borges Valeriano and Adão Felipe dos Santos
Advanced Farming Strategies Using NASA POWER Data in Peanut-Producing Regions without Surface Meteorological Stations
Reprinted from: *AgriEngineering* **2024**, *6*, 438-454, doi:10.3390/agriengineering6010027 . . . . . . **1**

Zhichong Wang, Yang Zhang, Tian Li, Joachim Müller and Xiongkui He
Visualization of Lidar-Based 3D Droplet Distribution Detection for Air-Assisted Spraying
Reprinted from: *AgriEngineering* **2023**, *5*, 1136-1146, doi:10.3390/agriengineering5030072 . . . . . **18**

Duc Anh Vu Trinh and Nguyen Truong Thinh
A Study of an Agricultural Indoor Robot for Harvesting Edible Bird Nests in Vietnam
Reprinted from: *AgriEngineering* **2024**, *6*, 113-134, doi:10.3390/agriengineering6010008 . . . . . . **29**

Vasiliki Kinigopoulou, Evangelos Hatzigiannakis, Stefanos Stefanou, Athanasios Guitonas and Efstathios K. Oikonomou
Utilization of Vermicompost Sludge Instead of Peat in Olive Tree Nurseries in the Frame of Circular Economy and Sustainable Development
Reprinted from: *AgriEngineering* **2023**, *5*, 1630-1643, doi:10.3390/agriengineering5030101 . . . . . **51**

Mino Sportelli, Christian Frasconi, Lorenzo Gagliardi, Marco Fontanelli, Michele Raffaelli and Massimo Sbrana et al.
Testing of Roller-Crimper-and-Undercutting-Blade-Equipped Prototype for Plants Termination
Reprinted from: *AgriEngineering* **2023**, *5*, 182-192, doi:10.3390/agriengineering5010013 . . . . . . **65**

Ted S. Kornecki and Corey M. Kichler
Influence of Recurrent Rolling/Crimping of a Cereal Rye/Crimson Clover Cover Crop on No-Till Bush Bean Yield
Reprinted from: *AgriEngineering* **2022**, *4*, 855-870, doi:10.3390/agriengineering4040055 . . . . . . **76**

Monika Vargová, Łukasz Bołoz, Miroslava Ťavodová and Richard Hnilica
Increasing the Durability of Tools for Forest Road Maintenance
Reprinted from: *AgriEngineering* **2023**, *5*, 566-580, doi:10.3390/agriengineering5010036 . . . . . . **92**

Pietro Toscano, Maurizio Cutini, Alex Filisetti, Elia Premoli, Maurizio Porcu and Nicola Catalano et al.
Workability Assessment of Different Stony Soils by Soil–Planter Interface Noise and Acceleration Measurement
Reprinted from: *AgriEngineering* **2022**, *4*, 1139-1152, doi:10.3390/agriengineering4040070 . . . . . **107**

Davide Gattamelata, Daniele Puri, Leonardo Vita and Mario Fargnoli
A Full Assistance System (FAS) for the Safe Use of the Tractor's Foldable Rollover Protective Structure (FROPS)
Reprinted from: *AgriEngineering* **2023**, *5*, 218-235, doi:10.3390/agriengineering5010015 . . . . . . **121**

Marcello Biocca, Daniele Pochi, Giancarlo Imperi and Pietro Gallo
Reduction in Atmospheric Particulate Matter by Green Hedges in a Wind Tunnel
Reprinted from: *AgriEngineering* **2024**, *6*, 228-239, doi:10.3390/agriengineering6010014 . . . . . . **139**

Daniele Pochi, Laura Fornaciari, Gennaro Vassalini, Renato Grilli and Roberto Fanigliulo
Levels of Whole-Body Vibrations Transmitted to the Driver of a Tractor Equipped with Self-Levelling Cab during Soil Primary Tillage
Reprinted from: *AgriEngineering* **2022**, *4*, 695-706, doi:10.3390/agriengineering4030044 . . . . . . **151**

Mino Sportelli, Sofia Matilde Luglio, Lisa Caturegli, Michel Pirchio, Simone Magni and Marco Volterrani et al.
Trampling Analysis of Autonomous Mowers: Implications on Garden Designs
Reprinted from: *AgriEngineering* **2022**, *4*, 592-605, doi:10.3390/agriengineering4030039 . . . . . . **163**

About the Editors

Marcello Biocca

Marcello Biocca (Orcid ID: "Biocca, M." 7003456986) is a senior researcher at the Engineering and Agri-Food Transformations Research Centre of CREA, Italy. He obtained a degree in Agricultural Sciences, and he is a Doctor of Philosophy (PhD) in Engineering for Energy and Environment.

He is carrying out research on the quality of agro-chemical applications, urban-forestry mechanization, and work safety. He has led research projects regarding the design and development of innovative machinery (also in collaboration with manufacturer companies) and assistance and demonstration programs. He has been a consultant in technical groups at the national level and has participated as an expert in national and European standardization technical tables.

He is a member of the Editorial Board of *AgriEngineering* (MDPI), and he has carried out refereeing activities for numerous scientific journals.

He is the author of about 230 scientific articles and publications, 2 patents, and 8 test reports and certification bulletins.

His interests are: agricultural engineering; crop protection; mechanization in oliviculture, urban forestry, and organic farming; work safety; human exposure to dust, noise, and vibrations.

Roberto Fanigliulo

Roberto Fanigliulo (ORCID ID: 0000-0002-4742-9408) graduated with a degree in Agricultural Science at the Università degli studi della Tuscia, Viterbo, Italy, where he also achieved the Ph.D. in Agricultural Mechanics. Since March 2008, he has been a full-time researcher at the Research Center for Engineering and Agro-Food Processing of the Council for Agricultural Research and Economics (CREA).

His research fields include: agricultural mechanization, prototype development; performance assessment of agricultural machinery (especially soil tillage and sowing machines), forestry equipment's, tires; verification of compliance of agricultural machinery safety requirements according to 2006/42/EC directive; study of airborne particulate from agricultural operations; study of bio-based lubricants and hydraulic fluids; experimental activity for evaluating their performance, aimed at introducing them in agricultural production systems to replace products of fossil origin.

He is the author of about 150 scientific articles and publications, 1 patent, and 180 test reports and certification bulletins.

Preface

Climate change, resource depletion, and evolving consumer demand impose the need for sustainable agricultural practices. Agricultural engineering and sustainable development are intrinsically linked, and all fields of agricultural engineering have great potential to improve global sustainability, from the development and use of innovative machinery to the latest application of digital and precision farming solutions. Agricultural engineering and mechanization play a key role in developing bio-based input, in waste processing and treatment to reduce greenhouse gas emissions, and in reducing and correctly distributing pesticides and fertilizers.

Agriculture, forestry, and food production must continue to provide safe and healthy food, feed, and fiber without consuming non-renewable resources or compromising the environment, ensuring economic viability and social and economic equity, and contributing to the achievement of sustainable development by the United Nations. objectives, the European Green Deal, and the Farm-to-Fork strategy.

This reprint, "Innovations in Agricultural Engineering and Mechanization for Sustainable Agriculture, Forestry, and Food Production", is a collection of cutting-edge research and advances in the field of agricultural engineering and mechanization, offering new perspectives and practical tools for navigating the complex realities of our time.

This volume delves into a wide range of topics, showcasing the originality of researchers from around the world. From the application of bio-based inputs and efficient distribution to advances in precision agriculture, robotics, and resource-saving technologies, the chapters within paint a comprehensive picture of the evolving landscape of sustainable food production.

This reprint is intended to be a valuable resource for a wide audience, including agricultural engineers and researchers who can gain insight into the latest research findings and explore potential avenues for further exploration and innovation, as well as policymakers and development professionals who can discover technologies and promising approaches.

Our hope is that this reprint will serve as a catalyst for further progress and inspire collective action towards a more sustainable and resilient agricultural sector.

Marcello Biocca and Roberto Fanigliulo
Editors

AgriEngineering

MDPI

Article

Advanced Farming Strategies Using NASA POWER Data in Peanut-Producing Regions without Surface Meteorological Stations

Thiago Orlando Costa Barboza [1,*], Marcelo Araújo Junqueira Ferraz [1], Cristiane Pilon [2], George Vellidis [2], Taynara Tuany Borges Valeriano [3] and Adão Felipe dos Santos [1,*]

1 Department of Agriculture, School of Agriculture Sciences of Lavras, Federal University of Lavras (UFLA), Lavras 37200-900, MG, Brazil; marcelo.ferraz1@estudante.ufla.br
2 Department of Crop and Soil Sciences, University of Georgia, Tifton, GA 31793, USA; cpilon@uga.edu (C.P.); yiorgos@uga.edu (G.V.)
3 Bayer CropScience, 51368 Leverkusen, Germany; taynarabvaleriano@gmail.com
* Correspondence: thiagocostaagro@gmail.com (T.O.C.B.); adao.felipe@ufla.br (A.F.d.S.)

Abstract: Understanding the impact of climate on peanut growth is crucial, given the importance of temperature in peanut to accumulate Growing Degree Days (GDD). Therefore, our study aimed to compare data sourced from the NASA POWER platform with information from surface weather stations to identify underlying climate variables associated with peanut maturity (PMI). Second, we sought to devise alternative methods for calculating GDD in peanut fields without nearby weather stations. We utilized four peanut production fields in the state of Georgia, USA, using the cultivar Georgia-06G. Weather data from surface stations located near peanut fields were obtained from the University of Georgia's weather stations. Corresponding data from the NASA POWER platform were downloaded by inputting the geographic coordinates of the weather stations. The climate variables included maximum and minimum temperatures, average temperature, solar radiation, surface pressure, relative humidity, and wind speed. We evaluated the platforms using Pearson correlation (r) analysis ($p < 0.05$), linear regression analysis, assessing coefficient of determination (R^2), root mean square error (RMSE), and Willmott index (d), as well as principal component analysis. Among the climate variables, maximum and minimum temperatures, average temperature, and solar radiation showed the highest R^2 values, along with low RMSE values. Conversely, wind speed and relative humidity exhibited lower correlation values with errors higher than those of the other variables. The grid size from the NASA POWER platform contributed to low model adjustments since the grid's extension is kilometric and can overlap areas. Despite this limitation, NASA POWER proves to be a potential tool for PMI monitoring. It should be especially helpful for growers who do not have surface weather stations near their farms.

Keywords: *Arachis hypogaea* L.; climate; weather data; peanut maturity (PMI); growing degree days

Citation: Barboza, T.O.C.; Ferraz, M.A.J.; Pilon, C.; Vellidis, G.; Valeriano, T.T.B.; dos Santos, A.F. Advanced Farming Strategies Using NASA POWER Data in Peanut-Producing Regions without Surface Meteorological Stations. *AgriEngineering* **2024**, *6*, 438–454. https://doi.org/10.3390/agriengineering6010027

Academic Editors: Marcello Biocca and Roberto Fanigliulo

Received: 10 October 2023
Revised: 15 January 2024
Accepted: 16 January 2024
Published: 20 February 2024

1. Introduction

Climate is extremely important in agricultural production, as a significant portion of production depends on specific climate conditions. In addition to water, temperature, relative humidity, solar radiation, and wind speed are factors that can affect production, along with the incidence of pests and diseases and soil microbiology [1]. Investigating climate change is necessary to adapt agricultural crop management, especially for plants in which growth, development, grain quality, and yield respond more sensitively to climatic variations [2].

Peanut (*Arachis hypogaea* L.) is among many crops that are affected by climate variations. Monitoring the climate has become of great importance to achieve gains in production. Peanut is produced worldwide, particularly in China (36%), India (13%), Nigeria (9%), and the United States (USA; 5%), with the state of Georgia accounting for 44% of the USA production [3].

Given the significance of peanut production, producers have increasingly utilized technologies to aid in monitoring and decision-making for peanut cultivation. In the USA, producers have employed the PeanutFarm (http://peanutfarm.org/ accessed on 2 February 2024) system to monitor peanut plant development. This system calculates the accumulation of degree days from meteorological stations distributed in various regions [4].

Despite the system's efficiency, a network of surface meteorological stations is necessary to supply climate data. An alternative approach involves the use of surface meteorological stations in a national network [5], data series obtained from mathematical models [6,7], and the use of orbital platforms.

Installing and monitoring meteorological stations is not easy, but it is necessary to understand the climatic conditions. Nevertheless, to monitor these climatic conditions and obtain accurate and precise values, an adequate number of meteorological stations is needed. In most of the countries, the number of weather stations is adequate. The World Meteorological Organization (WMO) recommends 6.3 stations per 100 km^2 [5,8]. An alternative for these countries is to use data from orbital platforms (satellites) with accurate models to monitor climatic changes. For instance, the use of temperature, precipitation, and relative humidity from the orbital platform is an alternative to creating evapotranspiration models to improve irrigation management and agricultural practices [9–11].

Several countries have adopted NASA POWER for climate monitoring, providing essential climatic information. The authors in [9] used the platform to estimate evapotranspiration in Lagunera, Mexico. In the semi-arid Mediterranean, the platform showed satisfactory adjustments to estimate daily evapotranspiration and improve irrigation methods [10]. The authors in [11] evaluated the accuracy and precision of NASA POWER climatic data in different climatic zones in Egypt, comparing it with surface weather stations. In Sicily, Italy, NASA POWER was used to estimate the reference evapotranspiration and apply it in regions that did not have weather stations to understand the impact of climate changes and improve agriculture [12].

One of the main orbital platforms for climate monitoring is NASA POWER, which collects information on a $1^\circ \times 1^\circ$ grid for solar radiation sources and a $\frac{1}{2}^\circ \times 5/8^\circ$ grid for climate data, enabling global climate monitoring. This tool has been applied to estimate corn productivity [5], leaf area, and productivity in soybeans [13] and develop models for identifying thermal stress [14].

Despite the use of grid data in various crops and for different purposes, there are no reports using these data to estimate peanut pod maturity, a crucial factor in determining grain productivity and quality. The maturation process of peanut pods depends on the accumulation of degree days by plants, with high temperatures accelerating growth and development, leading to faster maturation, while low temperatures can slow growth and delay maturity and, consequently, harvest time [15]. Monitoring pod maturity is crucial for farmers to improve their production and identify the optimum timing for inverting the peanut plants. This monitoring can be done using an orbital platform, such as NASA POWER, since the platform is online and publicly available.

However, there are limitations in using grid data. The data are collected using grids, and the grids have low spatial resolution, thus resulting in a loss of quality and detail richness. Errors in climate variable measurements can be encountered, such as precipitation and wind speed [5,16], affecting data quality and leading to erroneous analyses. Nevertheless, working with orbital data-collection tools that are publicly accessible can improve the understanding of climate changes and their effects on peanut cultivation and maturation, eliminating the need for meteorological stations near production areas.

Based on this, studies are required to investigate the reliability of data obtained from orbital platforms, as well as describe which variables are reliable for use in agriculture. Therefore, the objective of this work was to verify the applicability of remotely obtained

NASA POWER data to estimate peanut pod maturity and compare the data provided by NASA POWER with data obtained from surface meteorological stations.

2. Materials and Methods

2.1. Study Location

The fields used to assess the relation between pod maturity and climate data are in Georgia, USA. The region is classified as a subtropical humid climate (Cfa: temperate, without dry season and hot summer) with annual precipitation of 1346 mm [17]. Four fields were used (Figure 1) to evaluate the peanut maturity, with two fields being irrigated (Magnolia 2018 and Docia 2019) with center pivot and the other two fields being rainfed (Blaelock 2018 and Grand Canyon 2019). At each field, georeferenced points distanced 100 m apart were inserted, with 24 points (1 point/hectare) for Blaelock, Docia, and Magnolia and 12 points (1 point/hectare) for Grand Canyon to collect maturity samples.

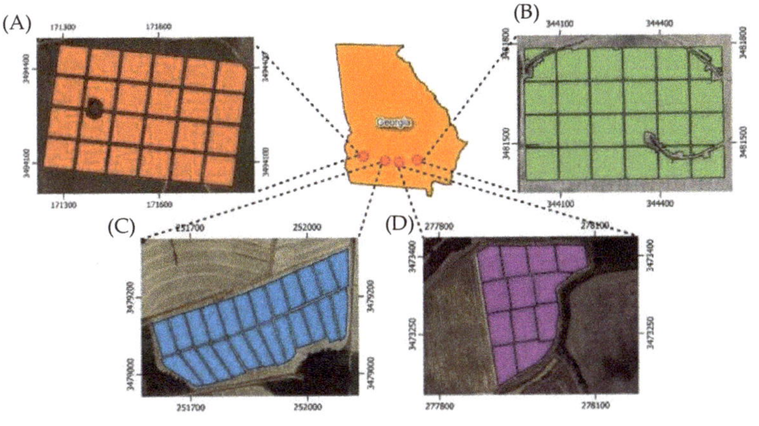

Figure 1. Peanut field production across various counties in Georgia, USA. (**A**) Magnolia (Ducker); (**B**) Blaelock (Coffee); (**C**) Docia (Tift); (**D**) Grand Canyon (Berrien). The red dots represent the location of each field in Georgia, and the color in each field represents the buffer delimiting the regions (polygons) of the collection of peanut to evaluate peanut pod maturity.

The fields were planted with the cultivar Georgia-06G, known for its dark green foliage and intermediate growth habit (Runner), with a production cycle of approximately 140 days [18]. Row spacing was 0.90 m. Planting was 5 June 2018 for Magnolia, 9 May 2019 for Docia, 11 June 2018, and 27 April 2019 for Grand Canyon. Each field was in a specific county: Magnolia in Ducker, Docia in Tift, Grand Canyon in Berrien, and Blaelock in Coffee.

2.2. Peanut Pod Maturity Evaluation

Peanut pods were collected from each field on different dates (Table 1). A total of 8 to 13 plants were collected around the georeferenced point (2 to 5 m), aiming for 200 pods per point (Figure 1). The collected plants were placed in a bag, identified, and transferred to the laboratory to detach the pods from the plants.

Table 1. Plant collection dates in days after sowing (DAS) from the fields for peanut pod maturity evaluation. Field counties are in parenthesis after each field name.

Fields	DAS
Blaelock (Coffee)	96; 104; 109; 116; 128; 135
Grand Canyon (Berrien)	103; 109; 117; 124; 131; 138; 145
Docia (Tift)	115; 122; 129; 139
Magnolia (Ducker)	96; 107; 117; 126; 135

Subsequently, the pods were pressure-washed, removing the exocarp and exposing the mesocarp [15], and classified according to peanut board maturity [19], which was adopted as the standard for evaluating maturity. The Peanut Maturity Index (PMI), ranging from 0 to 1, was obtained by summing up the brown and black columns of the maturation board and then dividing it by the total number of classified pods. Values close to 1 indicate mature pods, while values close to 0 indicate immature pods. However, under field conditions, optimal PMI values ranging from 0.7 to 0.75 were adopted to minimize quantitative and qualitative losses [15].

2.3. Climate Information

Meteorological stations located in Berrien (Alapaha), Tift (Tyty), Coffee (Douglas), and Dougherty (Ducker) were selected for collecting climatic data. These stations were the closest to the production fields described in Table 2.

Table 2. Distance (km) and elevation (m) of a surface weather station (SWS) of the production fields (PC) used to evaluate peanut pod maturity.

SWS	PC	Distance (km)	Elevation (m)
Berrien	Grand Canyon	9.0	82
Tift	Docia	10.5	113
Coffee	Blaelock	14.0	68
Dougherty	Magnolia	8.4	62

SWS—Surface weather stations; PC—production fields.

These stations near the collection fields for maturity assessment were selected to compose the analyses. Therefore, it was possible to carry out a comparative study between the climate variables predicted by NASA POWER and those observed by surface weather stations. The University of Georgia (UGA) Tifton Campus provided spreadsheets with weather data collected from the weather stations. However, due to the season collection system, climate variables data from Table 3 were recorded every 15 min. As the NASA POWER platform only provides daily data on climate variables, the data provided by the surface weather stations at UGA have been converted into a daily scale by calculating the average of the values of the climatic variables provided every 15 min. Thus, both stations were standard on a daily scale. In addition, data from the full years of 2018 and 2019 for the four weather stations were provided, creating the variables shown in Table 3.

Table 3. Agroclimatology variables obtained by fixed station and platform NASA POWER.

Climate Data	Unit of Measurement
Wind speed [1]	m/s
UR [2]	%
Tmax [3]	°C
Tmean [4]	°C
Tmin [5]	°C
SWN [6] (Qg)	MJ m^2 dia^{-1}
PS [7]	kPa

[1] speed of view at 2 m high; [2] humidity relative to 2 m high; [3] maximum temperature at 2 m high; [4] average temperature at 2 m tall; [5] minimum temperature at 2 m tall; [6] surface radiation incidences (solar radiation); [7] surface pressure.

On the NASA POWER platform (https://power.larc.nasa.gov/data-access-viewer/ accessed on 2 February 2024), the geographic coordinates of each surface weather station were inserted to collect the weather information described in Table 3. The platform provides daily information for each climate variable, with information being downloaded in CSV format for the full years of 2018 and 2019 from the four weather stations (Alapaha, Tyty, Douglas, and Ducker). The spatial resolution for the platform grid was $1° \times 1°$, which is approximately 12,347 km^2 for sources of primary solar radiation, whereas for weather data, regular grids of 0.5° × 0.625° of latitude/longitude, about 3850 km^2, were applied. The accuracy of the platform is adversely impacted by the use of large grids. For each variable, the root mean square error (RMSE) varies between 2.10, 3.15, and 3.10 °C for average, minimum, and maximum temperatures, respectively. Similarly, the RMSE values for wind speed, relative humidity, and surface pressure are 2.17 m/s, 12.06%, 2.87 kPa, and 6–12%, respectively [16,20].

The air temperature estimate was made using the Goddard Earth Observing System Global version 4 (GEOS-4), with an analysis interval of 3 h. Solar radiation data were obtained using the NASA International Satellite Cloud Climatology Project (ISCCP), and surface solar radiation was estimated using the ISCCP model [20,21].

With the collection of climate data from the two platforms, the comparison between the two forms of collection (terrestrial and orbital) was carried out, which indicated whether the NASA POWER data were accurate at estimating climate variables and consequently should be used in the estimation of peanut maturity anywhere around the globe.

2.4. Statistical Analysis of the Two Platforms

Climate data from the two platforms were combined into a general model, which considered the two years (2018 and 2019) of collection at all locations (Berrien, Coffee, Dougherty, and Tift). The climatic data from the two collection platforms were combined into a general model, which accounted for the two years (2018 and 2019). An additional segmentation was performed based on location, i.e., Berrien, Coffee, Dougherty, and Tift. This approach allowed for a specific analysis of each surface weather station throughout the two years of collection.

Initially, the full dataset was inserted into the exploration analysis using the boxplot, removing the values described as outliers by calculating the limits (inferior and superior). The weather data were analyzed using Pearson's correlation analysis ($p < 0.05$), and the graphs (heat maps) were plotted using the Jupyter platform with Python language. In addition to the correlation, linear regression analysis was performed for climate variables that showed a correlation coefficient above 0.8. Coefficient values between 0.67 and 1.0 [22] are generally considered to have a high correlation; however, for this study, the value of 0.8 was used to select the variables. For this analysis, the climatic variables of the surface weather stations were considered to be the dependent variables, whereas the independent variables were the variables provided by the NASA POWER platform. Exploratory analysis using boxplot and linear regression analysis was carried out using SAS© JMP pro 14 version 14.0.0 software, and regression graphs were created using Office Excel 2013 version 15.0.45 software. Subsequently, for the evaluation of the metrics of the models, the accuracy measurement of the RMSE (Equation (1)) and the determination coefficient (R^2) (Equation (3)) were used as a measure of precision.

Additionally, the calculation of (d), the Willmott index of conformity (1981), described in Equation (2), was carried out. The Willmott performance index (d) is a representation of the degree of error of the models, ranging from 0.0 to 1.0, with values close to 1.0 indicating a good match between the observed and predicted values [23].

$$RMSE = \sqrt{\frac{\sum_{i=1}^{n} \left(y_{obs} - y_{est}\right)^2}{n}} \tag{1}$$

$$d = 1 - \frac{\sum_{i=l}^{N} \left(Y_{obs_i} - Y_{est_i}\right)^2}{\sum_{i=l}^{N} \left(\left|Y_{est_i} - \overline{Y}\right| + \left|Y_{obs_i} - \overline{Y}\right|\right)} \tag{2}$$

$$R^2 = \frac{\sum_{i=l}^{N} (Yest_i - \overline{Y})^2}{(Yobs_i - \overline{Y})^2} \tag{3}$$

where *RMSE* is the square root of the average error, *d* is the Willmott concordance coefficient, and R^2 is the determination factor. y_{obs} is the observed value, y_{est} is the estimated value by the model, and *n* is the number of data points. $\overline{Y}$ the average value of the estimated variable.

Following the linear regression analysis, it was possible to show whether the NASA POWER platform is accurate and precise for estimating the climate variables found in Table 3. The variables were inserted in the principal component analysis (PCA). Thus, the dataset used was restricted to the periods of evaluation of maturity in the fields, and the PCA was carried out for each field (Berrien, Dougherty, Tift, and Coffee) and the Global model. Ultimately, the relationship between the PMI and climate variables can be identified, and one can select those variables that show the best results. With the PCA, the variables that best correlate with PMI can be selected, and the variables not showing a strong relationship can be excluded. PCA reduces the dimensionality of data while retaining as much information as possible. By transforming the data into principal components, it becomes feasible to concentrate on the directions that encompass the highest variability, therefore eliminating redundancies and emphasizing significant patterns. This technique is particularly valuable when dealing with datasets featuring numerous variables, aiding in the simplification of analysis and interpretation. The PCA was carried out using the software R, version 2023.06.2, and the package "factoextra", as shown in Figure 2.

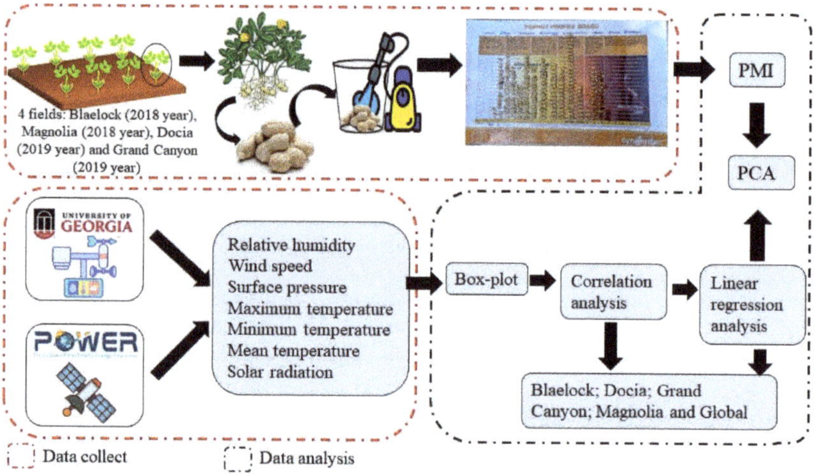

Figure 2. Evaluation steps for maturity classification and comparison between the NASA POWER platform and surface weather stations.

3. Results and Discussion

3.1. Correlation Analysis

The lowest correlation coefficients between climate data-collection platforms were found for wind speed (WS) and relative humidity (UR), 0.58 and 0.61 in the Global model, respectively (Figure 3E). For surface pressure (PS), solar radiation (Qg), maximum temperature, minimum temperature, and mean temperature, coefficient values were higher than 0.84 for the Global model (Figure 3E). The weak correlations for the UR and WS variables were due to interference in their localization, topography, and change in land use, which can cause errors in measurements when using grid data [11,16]. On the other hand, despite having a correlation value of 0.84, the PS perceived low reliability over weather stations in other studies [9], which also assessed the efficiency of NASA POWER.

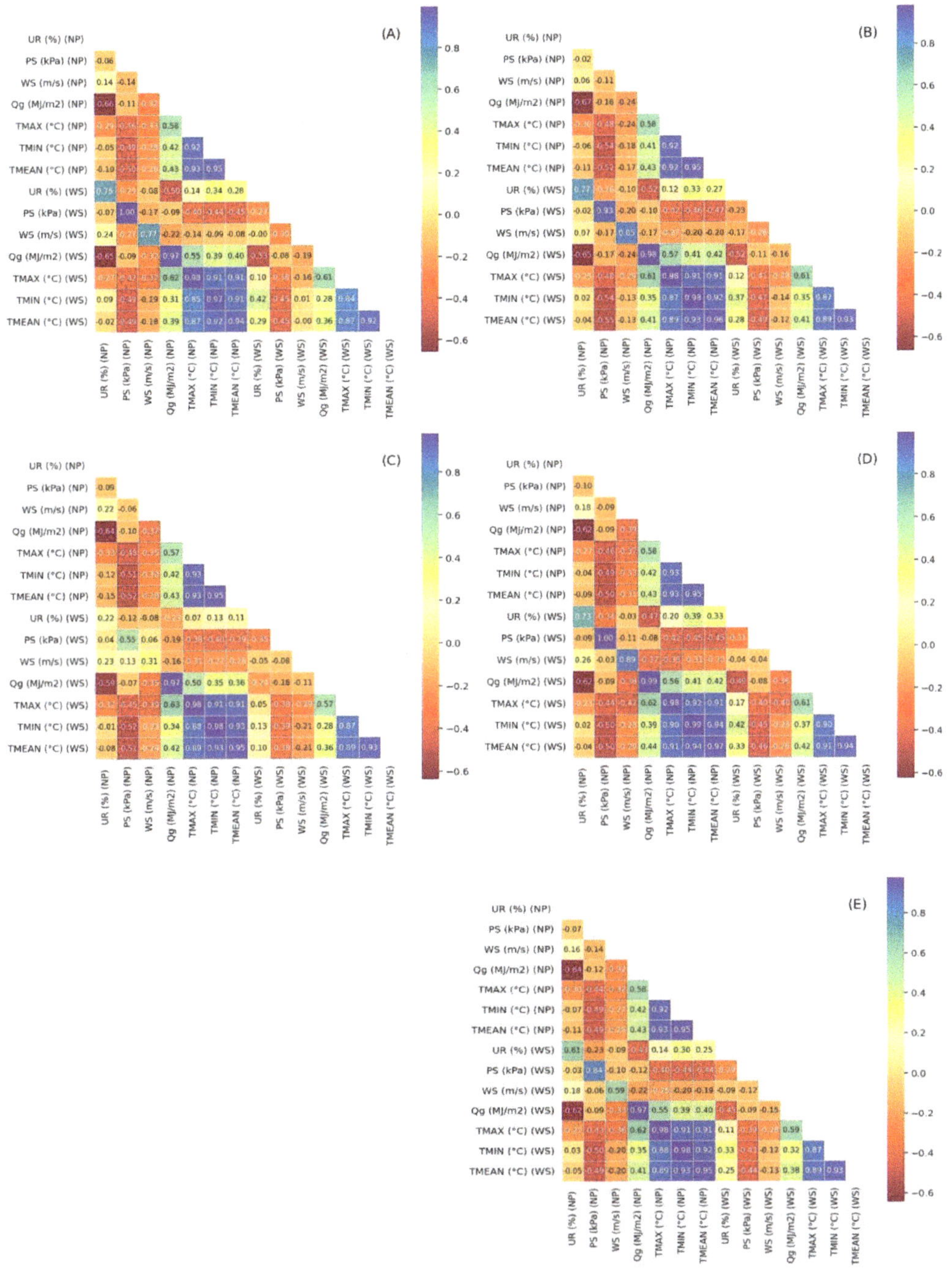

Figure 3. Correlation analysis for multiple variables from the locations Berrien (**A**), Coffe (**B**), Dougherty (**C**), Tift (**D**), and general model (**E**) for the two years (2018 and 2019).

When the correlation analysis was performed by location for Berrien, Coffee, and Tift (Figure 3A,B,D), the coefficients for the variables UR, WS, and PS were higher. Among these variables, WS changed from 0.59 in the global correlation (Figure 3E) to 0.85 and 0.89 for Coffee (Figure 3B) and Tift (Figure 3D), respectively. Although correlation values for WS had improved for these locations, when these variables (UR, WS, and PS) were analyzed individually in Dougherty (Figure 3C), the correlation values were lower. For maximum, minimum, and average temperatures and solar radiation, all locations showed similar results to the overall model.

The increase in the correlation coefficient can be attributed to the location conditions of each surface weather station since, in the NASA POWER data collection, there was no overlap of grids. Despite this, topography is one of the main factors affecting this relationship. The greater the elevation, the greater the errors described by the NASA POWER platform [12,24,25].

When the data were collectively analyzed to create a general model, the Pearson correlation values among all climatic variables showed a decrease (Figure 3E), i.e., Dougherty was the location that presented the lowest values for the Pearson correlation, with the greatest difference between the two platforms. When the locations were combined (Global), the correlation coefficient for the Global model decreased.

3.2. Regression Analysis

After conducting the correlation analysis, linear regression was performed using the climate variables, excluding relative humidity due to its Pearson correlation values being less than 0.8. The dependent variables (Y) were the surface weather station data, and the independent variable (X) was the NASA POWER data. Linear regression analysis was used to analyze the response of the variables and create a model to evaluate the precision (R^2) and accuracy (RMSE).

The linear regression analysis demonstrated that the precision values (R^2) for variable surface pressure (PS) were high for Berrien and Tift (Figure 4d,c), being $R^2 = 0.99$ and RMSE = 0.04 kPa for both locations. On the other hand, for the general model and the individual model for Coffee (Figure 4a,b), RMSE values were 0.25 and 0.18 kPa, and R^2 were 0.71 and 0.86, respectively. The results for Dougherty (Figure 4e) showed the largest variations between platforms for PS, which resulted in low R^2 (0.30) and high RMSE (0.36 kPa). In addition, the Willmott performance index (d) showed no variation in any of the analyses and local data for surface pressure.

Surface pressure was related to the displacement of water in the soil, causing the process of absorption of water and nutrients by the roots of plants. In addition, surface pressure is related to site topography, with regions with higher elevations showing lower PS, while higher PS values were observed for regions with lower elevations [25].

The WS had the highest variations in accuracy and precision levels in different locations. Dougherty (Figure 5e) was the county that showed the lowest adjustments of R^2 (0.09) and high RMSE (0.62 m/s), followed by the overall model (Figure 5a) with R^2 and RMSE values of 0.34 and 0.65 m/s, respectively. When there was separation by counties, the linear regression models for Berrien, Coffee, and Tift (Figure 5d,b,c) showed a greater adjustment of R^2, ranging from 0.59 to 0.79, and lower RMSE, from 0.36 to 0.42 m s^{-1}.

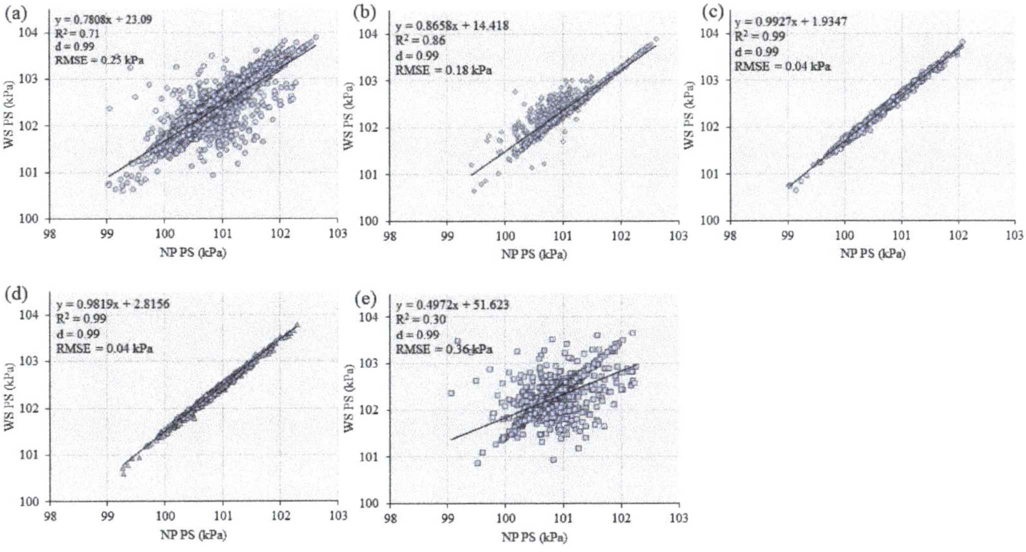

Figure 4. Linear regression analysis between NASA POWER (NP) and weather stations (WS) and metrics to evaluate the performance: determination coefficient (R^2), Root Mean Square Error (RMSE), and Willmott performance index (d) for surface pressure (PS). (**a**) General model; (**b**) Model for the Coffee region; (**c**) Model for the Tift region; (**d**) Model for the Berrien region; and (**e**) Model for the Dougherty region.

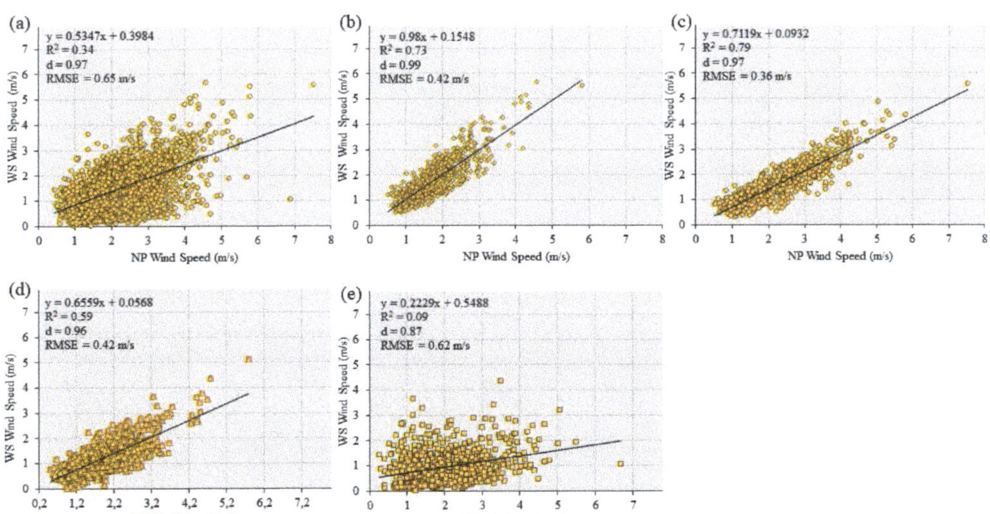

Figure 5. Linear regression analysis between NASA POWER (NP) and weather stations (WS) and metrics to evaluate the performance: determination coefficient (R^2), Root Mean Square Error (RMSE), and Willmott performance index (d) for Wind speed (WS). (**a**) General model; (**b**) Model for the Coffee region; (**c**) Model for the Tift region; (**d**) Model for the Berrien region; and (**e**) Model for the Dougherty region.

Despite the variations observed in the R^2 and RMSE values for the different models, in comparative studies between the NASA POWER platform and the national network stations in Brazil, R^2 values ranging from 0.09 to 0.16 and RMSE from 0.93 to 1.82 m s^{-1} were observed [5]. As reported by [9], models had RMSE values ranging from 0.92 to 1.63 m s^{-1} and R^2 between 0.19 to 0.52, which were close to the variations observed in this study. In addition, for most locations, the data points shown in Figure 5a–e are more concentrated around 1.2 to 3.2 m s^{-1}, being more dispersed outside this range.

The variability observed in the measurement of wind speed (WS) is associated with how these data are captured by the sensor and calculated through mathematical models. The mathematical models used in the NASA POWER platform are the Modern-Era Retrospective Analysis for Research and Applications 2 (MERRA-2), which calculates the speed and direction of the wind, and the results are compared with NASA's weather stations, with RMSE values of up to 2.47 m s^{-1} [20].

In agriculture, wind speed is an important factor for crop evaporation. In defining the planting window for peanut crops, plantings occurring in mid-May showed greater evapotranspiration values that increased leaf area [26]. This fact, which coincides with the season of the highest wind speed values (Figure S2), was recorded in the spring season at the beginning of sowing in Georgia, USA.

All models for daily solar radiation (Qg) showed R^2 adjustments above 0.94. Higher R^2 and lower RMSE values were observed for the general model and Berrien (Figure 6a,d). In Tift, the best adjustments for Qg were observed with R^2 = 0.97 and RMSE = 1.12 MJ m^2/day (Figure 6c). The variation in RMSE values was 0.48 MJ m^2/day between the highest (Figure 6a) and the lowest (Figure 6c) values observed, and there were no variations for any model in the Willmott performance index values (d = 0.99). The errors found in this study can be considered to be low since the error values for Qg estimated by NASA POWER range from 2.73 to 3.41 MJ m^2/day [5,9].

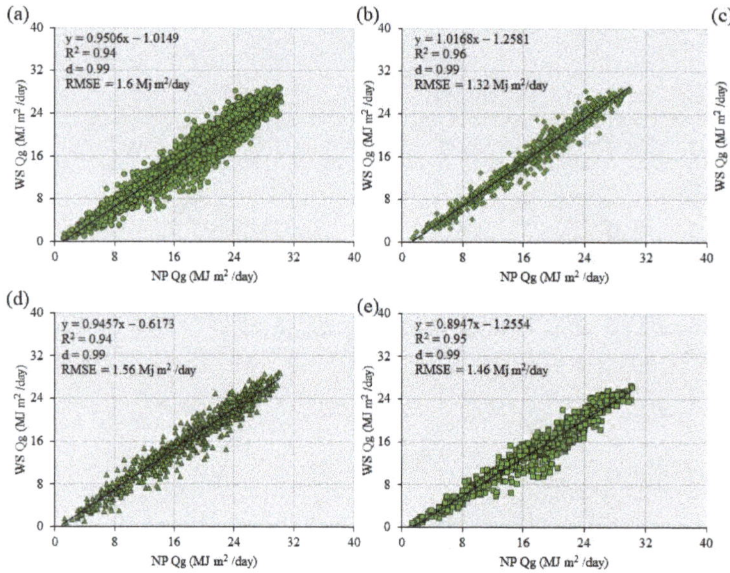

Figure 6. Linear regression analysis between NASA POWER (NP) and weather stations (WS) and metrics to evaluate the performance: determination coefficient (R^2), Root Mean Square Error (RMSE), and Willmott performance index (d) for Solar radiation (Qg). (**a**) General model; (**b**) Model for the Coffee region; (**c**) Model for the Tift region; (**d**) Model for the Berrien region; and (**e**) Model for the Dougherty region.

Solar radiation is a parameter dependent on weather conditions, and the presence of clouds makes its analysis process more challenging, leading to errors [5] and, therefore, decreasing accuracy and precision. However, the models showed satisfactory values of accuracy and precision, even when data from all sites from the two years (Global) and when separated by location were used, demonstrating the platform's precision in estimating Qg at any time of the year. For the calculation of solar radiation, the mathematical model Global Energy and Water Exchanges (GEWEX) used by the NASA POWER platform features more satellites that capture information about cloud coverage, as well as other satellites to provide the temperature and gas data in the atmosphere. These satellites provide information to the radiative transfer models for the correction of the effects of these constituents on the estimation of solar radiation [20].

Furthermore, Qg is a temperature-dependent parameter that can influence both air and soil humidity. The seasons with the highest mean temperature and relative humidity (see Supplementary Materials)—spring and summer—recorded the highest Qg values. This period of elevated Qg corresponds to the peanut-growing season in Georgia. Therefore, late-summer seedings outside the optimal planting window may result in reduced productivity due to changes in climate conditions, particularly in temperature and solar radiation. The decrease in solar radiation and temperature decreases leaf photosynthetic rates, leading to a reduction in plant growth, biomass accumulation, and decreased productivity [26].

For maximum temperature, the observed R^2 values ranged from 0.95 for the general model, Coffee and Tift (Figure 7a–c) to 0.96 for Berrien and Dougherty (Figure 7d,e). The lowest RMSE values were 1.53 °C for Berrien (Figure 7d), and the highest of 1.63 °C (Figure 7e) was for Dougherty, with a variation of 0.1 °C between maximum and minimum RMSE observed in these two locations. The Willmott performance index (d) was 0.99, showing no significant difference for any of the tested models and for maximum, mean, and minimum temperatures.

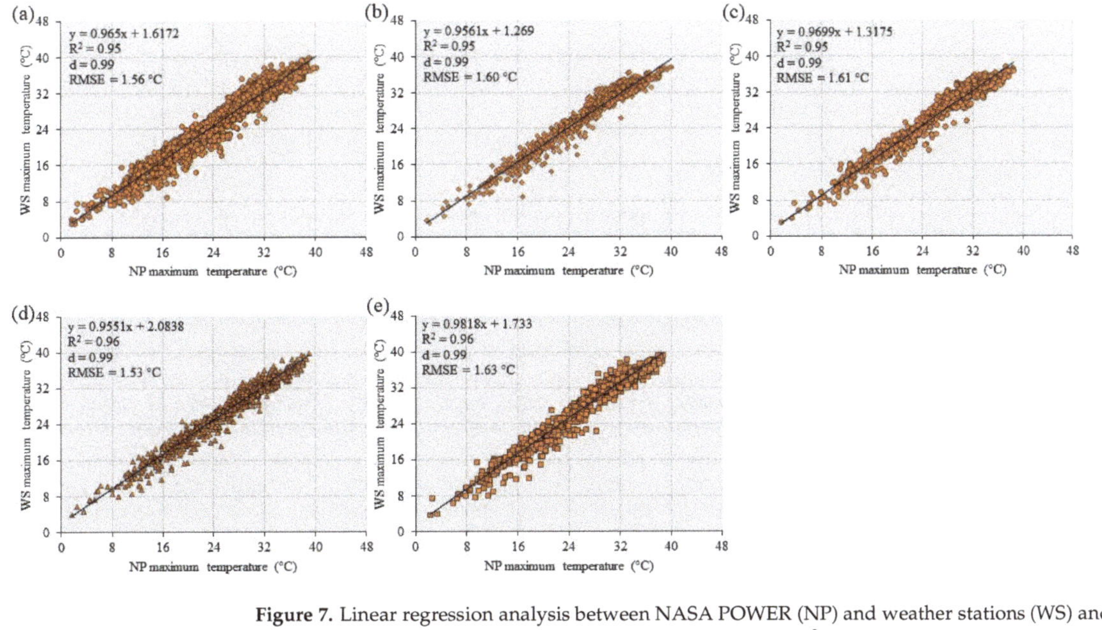

Figure 7. Linear regression analysis between NASA POWER (NP) and weather stations (WS) and metrics to evaluate the performance: determination coefficient (R^2), Root Mean Square Error (RMSE), and Willmott performance index (d) for maximum temperature. (**a**) General model; (**b**) Model for the Coffee region; (**c**) Model for the Tift region; (**d**) Model for the Berrien region; and (**e**) Model for the Dougherty region.

Both data-collection platforms recorded negative values for the minimum temperature variable. Despite this, satisfactory adjustments were obtained from the models described by the precision values (R^2) of 0.96 for the general model, Coffee and Dougherty (Figure 8a,b,e), 0.97 for Tift (Figure 8c), and 0.94 for Berrien (Figure 8d). With regard to accuracy, Tift was the region with the lowest values of 1.24 °C (RMSE), whereas Berrien was the one with the highest levels of 1.84 °C (Figure 8c,d).

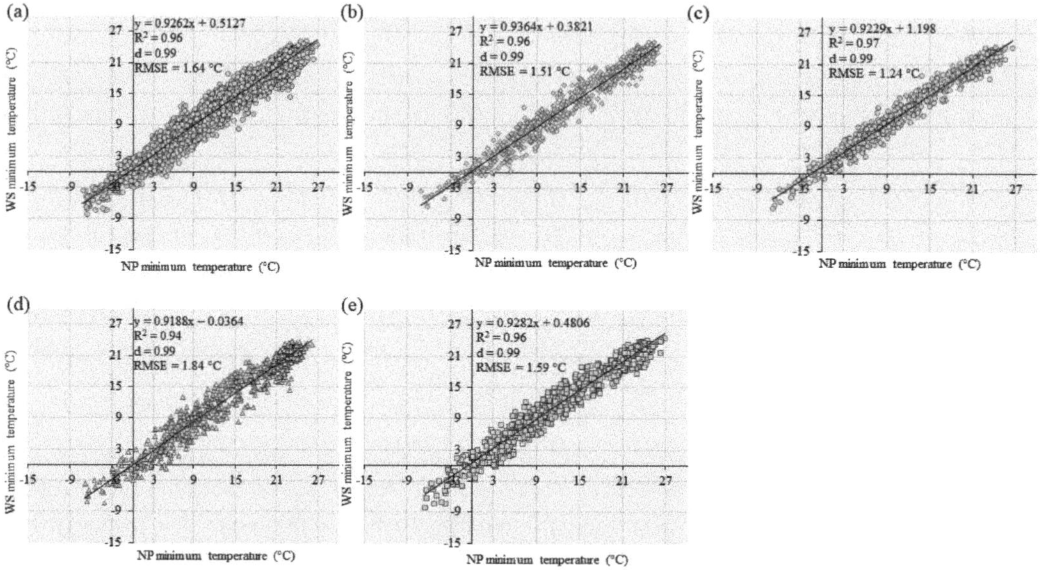

Figure 8. Linear regression analysis between NASA POWER (NP) and weather stations (WS) and metrics to evaluate the performance: determination coefficient (R^2), Root Mean Square Error (RMSE), and Willmott performance index (d) for minimum temperature. (**a**) General model; (**b**) Model for the Coffee region; (**c**) Model for the Tift region; (**d**) Model for the Berrien region; and (**e**) Model for the Dougherty region.

The linear regression models exhibited the least adjustments for average air temperature. In the general model, Coffee and Dougherty (Figure 9a,b,e) demonstrated an accuracy value of R^2 = 0.91, while Tift exhibited an R^2 = 0.93 (Figure 9c). Conversely, Berrien (Figure 9d) displayed the lowest R^2 = 0.89. The RMSE varied from 2.44 °C for Tift (Figure 9c) to 3.43 °C for Berrien (Figure 9d).

For maximum and minimum air temperatures, low variations in the data distribution in the regression line were recorded, resulting in satisfactory adjustments for the linear regression models. On the other hand, for the average air temperature, data points were more scattered from the adjustment line of the linear regression model. Such data dispersion can be attributed to the way data are collected from the two platforms. Although one platform recorded a positive value, the other platform recorded a negative value for the same date, affecting the fit of the models and, consequently, the parameters of accuracy (RMSE) and precision (R^2). The dispersion error persists due to the computational approach of the NASA POWER platform models. Given that the data are presented in grids (0.5° × 0.625°) by the MERRA-2 model, the substantial extent of the grid, exceeding 50 km, may introduce errors in calculations. This discrepancy is particularly notable in locations where weather conditions diverge from those observed by nearby field weather stations. Unlike weather stations, which gather more precise information from specific locations, the grid-based approach may aggregate diverse conditions, contributing to inaccuracies.

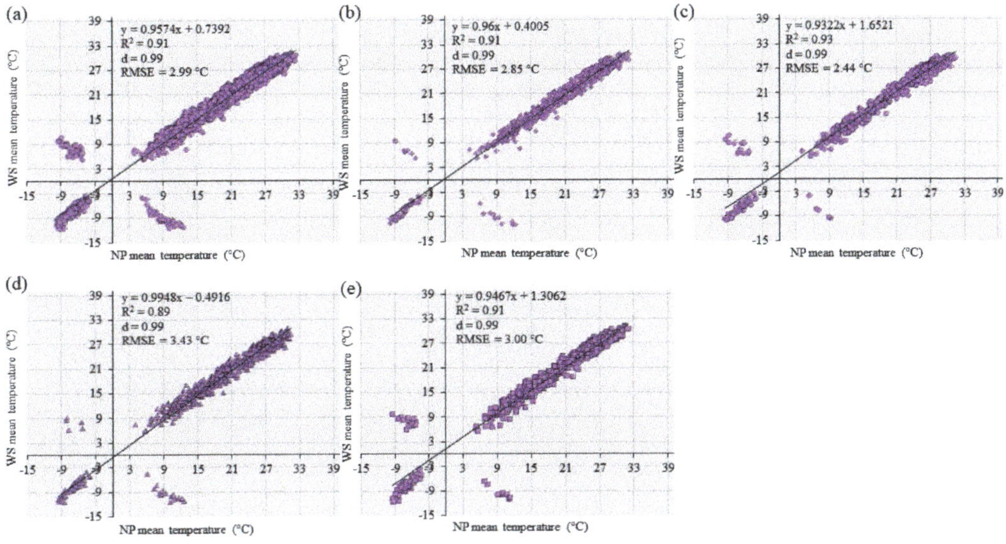

Figure 9. Linear regression analysis between NASA POWER (NP) and weather stations (WS) and metrics to evaluate the performance: determination coefficient (R^2), Root Mean Square Error (RMSE), and Willmott performance index (d) for mean temperature. (**a**) General model; (**b**) Model for the Coffee region; (**c**) Model for the Tift region; (**d**) Model for the Berrien region; and (**e**) Model for the Dougherty region.

In studies with the NASA POWER platform and surface ground stations, R^2 values ranging from 0.84 to 0.95 and RMSE from 1.29 °C to 3.67 °C for maximum, minimum, and average air temperature were observed [9].

In different scenarios, while working with a network of stations in Brazil, the RMSE can range from 2.64 °C to 2.83 °C, with corresponding R^2 values varying between 0.68 and 0.65 for mean temperature [5]. Similarly, for maximum temperature, the R^2 values range from 0.08 to 0.63, and for minimum temperature, they vary from 0.08 to 0.85 [27]. On the other hand, the NASA POWER program reported that errors (RMSE) of 2.10 °C were found in models (MERRA-2) when estimating average air temperature [20,21]. Conversely, in this study, great adjustments were observed for the models, described by high values of R^2 and low RMSE for maximum and minimum air temperatures. The topographic conditions and soil usage are crucial factors for characterizing the climate of a site. In Georgia, varying temperature ranges have been recorded, influenced by the region within the state. For northern regions near Tennessee, temperature ranged from 3.0 to 5.9 °C in January, while in the southern regions near Florida, temperature variations were from 8.0 to 14.9 °C for the same month of the year [28,29].

Determining the optimal sowing timing for crops is crucial to securing favorable climate conditions for cultivation. Depending on the seeding season, temperature significantly impacts dry-matter production, leaf growth, and peanut germination. However, the rapid initial growth, influenced by the elevated temperatures in June (27 and 33 °C), plays a significant role in plant-stand establishment and, consequently, production [30].

The peanut plant is substantially impacted by temperature, which can affect both the maturing process and the overall quality of harvested pods. This becomes apparent when employing agrometeorological indices, such as accumulated degrees days (AGD), for evaluating PMI. This index has already been used in various studies since the responses observed between maturity and AGD are satisfactory [4,15,31]. For the calculation, in addition to maximum and minimum air temperatures, the base temperature of 13.5 °C for peanuts is required [26,30]. Another important aspect is that temperature can also influence other climate variables, therefore altering how the crop is managed.

Temperature is an important factor from a climate point of view, and climate variables such as relative humidity and solar radiation are influenced by their changes. In the management of agricultural crops, these climate variables collaborate to obtain high productivity. However, for the monitoring of weather conditions, grid platforms such as NASA POWER are low-cost and feasible tools that can be applied to the analysis of local weather conditions. Such platforms show similar results when compared to surface weather stations, mainly for the variables of surface pressure, maximum, minimum, and average air temperature, and solar radiation (Figures 4–9).

3.3. *Principal Component Analysis (PCA)*

Climate data are relevant from an agronomic point of view, as they can interfere with the productivity of agricultural crops. However, to assess the climatological variables that are most related to maturity in peanut, the principal component analysis (PCA) shown in Figure 10 was performed. In the PCA analysis, the variables surface pressure, relative humidity, maximum, minimum, and average temperatures, as well as wind speed and solar radiation, were included, and the relationship of climate variables with maturity (PMI) could be observed.

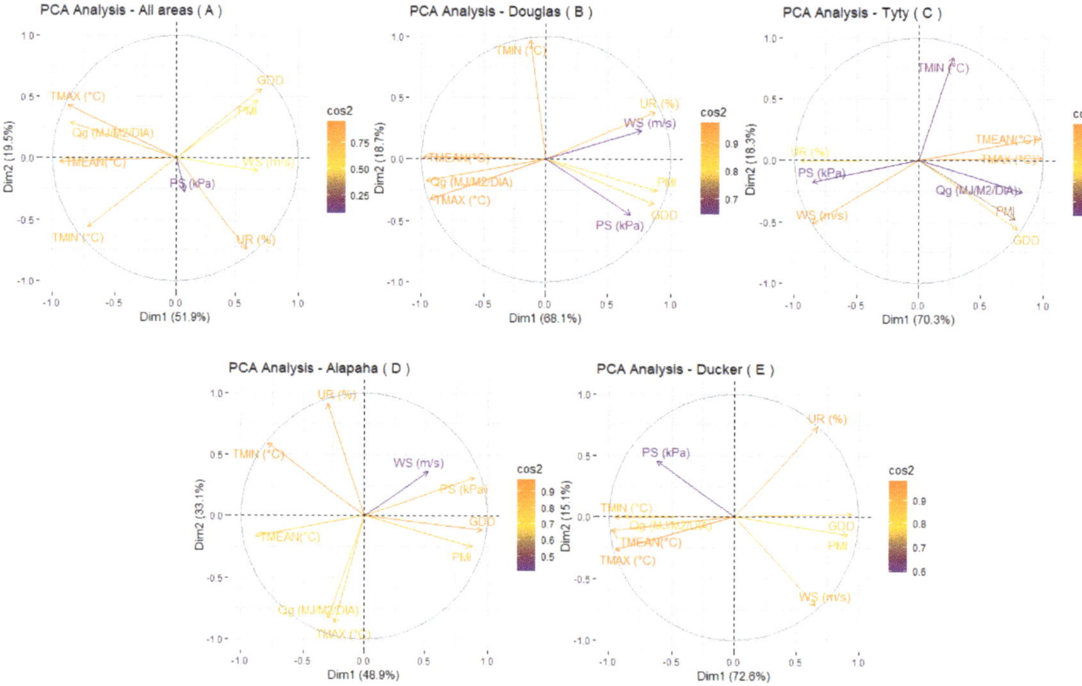

Figure 10. Principal Component Analysis (PCA) for each region and Global model. (**A**) represents the PCA of the general model; (**B**) represents the Coffee region PCA; (**C**) represents the Tift region PCA; (**D**) represents the Berrien region; and (**E**) represents the Dougherty region. PS: surface pressure; WS: wind speed; UR: relative humidity; QG: solar radiation; Tmax: maximum temperature; Tmin: minimum temperature; Tmean: average temperature.

The general model (Figure 10A), considering the two PCA components, accounted for 71.8% of the data variability. For the analysis by region, the sum of components 1 and 2 accounted for 82%, 87,1%, 87.7%, and 88.2% of the overall data variability for

Berrien (Figure 10D), Coffee (Figure 10B), Dougherty (Figure 10E), and Tift (Figure 10C), respectively. These results demonstrated an increase in PCA's ability to respond to data variability. Furthermore, it was observed that Tmin, Tmax, Tmean, and Qg follow an opposite trend than PMI, demonstrating that such climate variables affect PMI values, i.e., as temperature or solar radiation increase, PMI value decreases, consequently decreasing the productivity and quality of harvested pods.

Thus, the definition of the growing season is an essential factor for the development of peanut plants. The sowing window for peanuts begins around 10 April and lasts until the beginning of July. In this seeding season, the ideal climatic conditions for peanut are found, with the optimal temperature for growth being 27.5 °C, which can vary from 29 °C to 33 °C [31]. These variables that may interfere with the development of peanut plants can be estimated (maximum, minimum, average air temperatures, and solar radiation) using the NASA POWER platform (Figures 6–9) with high accuracy and precision. However, it should be observed whether peanut-growing areas do not overlap due to the low spatial resolution of $0.5^\circ \times 0.625^\circ$ of latitude and longitude (approximately 55.6×69.4 km) for meteorological data and $1^\circ \times 1^\circ$ of latitude and longitude for solar radiation data (approximately 111×111 km) of grid data.

4. Conclusions

In conclusion, this study has successfully demonstrated the viability of utilizing NASA POWER data for monitoring climatic conditions, showcasing strong correlations between maximum, minimum, and average air temperatures, as well as solar radiation when compared to surface weather stations. Notably, these variables exhibited significant relationships with peanut pod maturity, as highlighted in the PCA analysis.

Despite the promising results, certain limitations were identified, particularly for wind speed, which displayed challenges in achieving accurate and precise adjustments in linear regression models. This discrepancy can be attributed to the difference in measurement heights between NASA POWER (50 m) and weather stations (2 m), impacting the overall fit, especially in the Dougherty region.

However, the NASA POWER platform emerges as a valuable tool for climatic monitoring. Farmers can leverage this platform to gain insights into crop behavior across diverse climates. The broader application extends to areas without surface weather stations, enabling accurate monitoring and providing a useful tool for understanding climatic changes.

Although the platform proves instrumental, it is essential to acknowledge its limitations, such as low spatial resolution with grids larger than 50 km for weather data and more than 100 km for solar radiation. This may introduce restrictions and potential interference in data analysis. For instance, relative humidity showed a correlation below 0.8 when compared to surface weather stations, suggesting caution in its interpretation, especially in regions with monitoring stations reporting errors.

In terms of innovation, this work proposes a groundbreaking approach to monitoring climatic conditions on farms using publicly accessible orbital platforms. The NASA POWER platform stands out as an excellent resource, empowering farmers in peanut fields to calculate indices and effectively monitor climate parameters. Moreover, regions lacking surface weather stations can rely on the orbital platform to access crucial climatic information, contributing to more informed agricultural practices and climate monitoring on a broader scale.

Supplementary Materials: The following supporting information can be downloaded at https://www.mdpi.com/article/10.3390/agriengineering6010027/s1. Display of boxplot graphs illustrating the variability throughout the seasons of the year between the NASA POWER platform and surface weather stations.

Author Contributions: For the research article, the authors' contributions are as follows: Conceptualization, C.P., A.F.d.S. and G.V.; Methodology A.F.d.S. and T.T.B.V.; Validation T.O.C.B. and M.A.J.F.; Formal Analysis A.F.d.S., T.O.C.B. and M.A.J.F.; Writing—Review and Editing T.O.C.B., A.F.d.S. and C.P.; Visualization T.T.B.V. and G.V.; Project Administration A.F.d.S.; Writing—original draft preparation T.O.C.B. All authors have read and agreed to the published version of the manuscript.

Funding: This research was funded by the Minas Gerais State Research Support Foundation (FAPEMIG), process number APQ-00219-21.

Data Availability Statement: Data are contained within the article.

Acknowledgments: The authors acknowledge that the Minas Gerais State Research Support Foundation (FAPEMIG) funded the project and the Coordination of Superior Level Staff Improvement (CAPES) by the financial support of students. The Federal University of Lavras (UFLA) made the laboratory available and contributed to the structure. The University of Georgia (UGA) helped during the analysis, revisions, and collaboration with the data.

Conflicts of Interest: The authors declare that the research was conducted in the absence of any commercial or financial relationships that could be construed as a potential conflict of interest.

References

1. Jeger, M.J. The impact of climate change on disease in wild plant populations and communities. *Plant Pathol.* **2021**, *71*, 111–130. [CrossRef]
2. Wei, S.; Li, K.; Yang, Y.; Wang, C.; Liu, C.; Zhang, J. Comprehensive climatic suitability evaluation of peanut in Huang-Huai-Hai region under the background of climate change. *Sci. Rep.* **2022**, *12*, 11350. [CrossRef]
3. USDA—United States Department of Agriculture. USDA. 2020. Available online: https://ipad.fas.usda.gov/cropexplorer/cropview/commodityView.aspx?cropid=2221000 (accessed on 18 July 2023).
4. Rowland, D. *Peanut Field Agronomic Resource Manager (PeanutFARM)*; University of Florida, Agronomy Department: Gainesville, FL, USA. Available online: http://peanutfarm.org/overviews (accessed on 30 October 2023).
5. Duarte, Y.C.N.; Sentelhas, P.C. NASA/POWER and DailyGridded weather datasets—How good they are for estimating maize yields in Brazil? *Int. J. Biometeorol.* **2019**, *64*, 319–329, Erratum in *Int. J. Biometeorol.* **2020**, *64*, 331–332. [CrossRef]
6. Voyant, C.; Muselli, M.; Paoli, C.; Nivet, M.-L. Numerical weather prediction (NWP) and hybrid ARMA/ANN model to predict global radiation. *Energy* **2012**, *39*, 341–355. [CrossRef]
7. Bengtsson, L.; Andrae, U.; Aspelien, T.; Batrak, Y.; Calvo, J.; de Rooy, W.; Gleeson, E.; Hansen-Sass, B.; Homleid, M.; Hortal, M.; et al. The HARMONIE-AROME model configuration in the ALADIN-HIRLAM NWP system. *Mon. Weather. Rev.* **2017**, *145*, 1919–1935. [CrossRef]
8. Plummer, N.; Allsopp, T.; Lopez, J.A. *Guidelines on Climate Observation Networks and Systems*; WMO/TD No. 1185; World Meteorological Organization: Geneva, Switzerland, 2003.
9. Jiménez, S.I.; Bustamante, W.O.; Ibarra, M.A.I.; Pablo, M.J.P. Analysis of the NASA-POWER system for estimating reference evapotranspiration in the Comarca Lagunera, Mexico. *Ing. Agr. Biosist.* **2021**, *13*, 201–226. [CrossRef]
10. Rodrigues, G.C.; Braga, R.P. Estimation of Daily Reference Evapotranspiration from NASA POWER Reanalysis Products in a Hot Summer Mediterranean Climate. *Agronomy* **2021**, *11*, 2077. [CrossRef]
11. Aboelkhair, H.; Morsy, M.; El Afandi, G. Assessment of agroclimatology NASA POWER reanalysis datasets for temperature types and relative humidity at 2 m against ground observations over Egypt. *Adv. Space Res.* **2019**, *64*, 129–142. [CrossRef]
12. Negm, A.; Jabro, J.; Provenzano, G. Assessing the suitability of American national Aeronautics and Space Adminstration (NASA) agro-climatology archive to predict daily meteorological variables and reference evapotranspiration in Sicily, Italy. *Agric. For. Meteorol.* **2017**, *244*, 111–121. [CrossRef]
13. Gaso, D.V.; de Wit, A.; de Bruin, S.; Puntel, L.A.; Berger, A.G.; Kooistra, L. Efficiency of assimilating leaf area index into a soybean model to assess within-field yield variability. *Eur. J. Agron.* **2023**, *143*, 126718. [CrossRef]
14. Palmero, F.; Carcedo, A.J.; Haro, R.J.; Bigatton, E.D.; Salvagiotti, F.; Ciampitti, I.A. Modeling drought stress impacts under current and future climate for peanut in the semiarid pampas region of Argentina. *Field Crop. Res.* **2022**, *286*, 108615. [CrossRef]
15. Rowland, D.L.; Sorensen, R.B.; Butts, C.L.; Faircloth, W.H. Determination of Maturity and Degree Day Indices and their Success in Predicting Peanut Maturity. *Peanut Sci.* **2006**, *33*, 125–136. [CrossRef]
16. Maldonado, W.; Valeriano, T.T.B.; Rolim, G.d.S. EVAPO: A smartphone application to estimate potential evapotranspiration using cloud gridded meteorological data from NASA-POWER system. *Comput. Electron. Agric.* **2019**, *156*, 187–192. [CrossRef]
17. Beck, H.E.; Zimmermann, N.E.; McVicar, T.R.; Vergopolan, N.; Berg, A.; Wood, E.F. Present and Future Köppen-Geiger Climate Classification Maps at 1-km Resolution. *Scientific Data* **2018**, *5*, 180214. [CrossRef]
18. Branch, W.D. Registration of 'Georgia-06G' peanut. *J. Plant Regist.* **2007**, *1*, 120. [CrossRef]
19. Williams, E.J.; Drexler, J.S. A Non-Destructive Method for Determining Peanut Pod Maturity. *Peanut Sci.* **1981**, *8*, 134–141. [CrossRef]

20. Stackhouse, P. *Metodologia: Fonte de Dados*; NASA Docs; NASA: Washington, DC, USA, 2022. Available online: https://power.larc.nasa.gov/docs/methodology/data/sources/ (accessed on 30 October 2023).
21. Bai, J.; Chen, X.; Dobermann, A.; Yang, H.; Cassman, K.G.; Zhang, F. Evaluation of NASA satellite- and model-derived weather data for simulation of maize yield potential in China. *Agron. J.* **2010**, *102*, 9–16. [CrossRef]
22. Taylor, R. Interpretation of the Correlation Coefficient: A Basic Review. *J. Diagn. Med. Sonogr.* **1990**, *6*, 35–39. [CrossRef]
23. Willmott, C.J.; Ackleson, S.G.; Davis, R.E.; Feddema, J.J.; Klink, K.M.; Legates, D.R.; O'Donnell, J.; Rowe, C.M. Statistics for the evaluation and comparison of models. *J. Geophys. Res.* **1985**, *90*, 8995–9005. [CrossRef]
24. Valeriano, T.T.B.; Rolim, G.d.S.; Bispo, R.C.; de Moraes, J.R.d.S.C.; Aparecido, L.E.d.O. Evaluation of air temperature and rainfall from ECMWF and NASA gridded data for southeastern Brazil. *Theor. Appl. Climatol.* **2019**, *137*, 1925–1938. [CrossRef]
25. Richardson, M.I.; Newman, C.E. On the relationship between surface pressure, terrain elevation, and air temperature. Pate I: The large diurnal surface pressure range at Gale Crater, Mars and its origin due to lateral hydrostatic adjustment. *Planet. Space Sci.* **2018**, *164*, 132–157. [CrossRef]
26. Zhang, G.; Leclerc, M.; Singh, N.; Tubbs, R.; Monfort, W. Impact of planting date on CO_2 fluxes, evapotranspiration and water-use efficiency in peanut using the eddy-covariance technique. *Agric. For. Meteorol.* **2022**, *326*, 109163. [CrossRef]
27. Aguiar, J.T.; Junior Lobo, M. Reliability and discrepancies of rainfall and temperatures from remote sensing and Brazilian ground weather stations. *Remote Sens. Appl. Soc. Environ.* **2020**, *18*, 100301.
28. Soulé, P.T. Some Spatial Aspects of Southeastern United States Climatology. *J. Geogr.* **1998**, *97*, 142–150. [CrossRef]
29. Binita, K.C.; Shepherd, M.J.; Gaither, C.J. Climate change vulnerability assessment in Georgia. *Appl. Geogr.* **2015**, *62*, 62–74.
30. Virk, G.; Pilon, C.; Snider, J.L. Impact of First True Leaf Photosynthetic Efficiency on Peanut Plant Growth under Different Early-Season Temperature Conditions. *Peanut Sci.* **2019**, *46*, 162–173. [CrossRef]
31. Russelle, M.P.; Wilhelm, W.W.; Olson, R.A.; Power, J.F. Growth Analysis Based on Degree Days. *Crop. Sci.* **1984**, *24*, 28–32. [CrossRef]

AgriEngineering

MDPI

Article

Visualization of Lidar-Based 3D Droplet Distribution Detection for Air-Assisted Spraying

Zhichong Wang [1,2], Yang Zhang [1], Tian Li [2,3,4], Joachim Müller [1] and Xiongkui He [2,3,4,*]

1 Tropics and Subtropics Group, Institute of Agricultural Engineering, University of Hohenheim, 70599 Stuttgart, Germany; z.wang@uni-hohenheim.de (Z.W.); yang.zhang@uni-hohenheim.de (Y.Z.); joachim.mueller@uni-hohenheim.de (J.M.)
2 College of Agricultural Unmanned System, China Agricultural University, Beijing 100193, China; cault@cau.edu.cn
3 College of Science, China Agricultural University, Beijing 100193, China
4 Centre for Chemicals Application Technology, China Agricultural University, Beijing 100193, China
* Correspondence: xiongkui@cau.edu.cn

Abstract: Air-assisted spraying is a commonly used spraying method for orchard plant protection operations. However, its spraying parameters have complex effects on droplet distribution. The lack of large-scale 3D droplet density distribution measurement methods of equipment has limited the optimization of spraying parameters. Therefore, there is a need to develop a method that can quickly obtain 3D droplet distribution. In this study, a 2D LiDAR was used to quickly scan moving droplets in the air, and a test method that can obtain the visualization of 3D droplet distribution was constructed by using the traveling mode of the machine perpendicular to the scanning plane. The 3D droplet distribution at different positions of the nozzle installed in the air-assisted system was tested at different fan rotation speeds, and the methods for signal processing, point cloud noise reduction, and point cloud division for 2D LiDAR were developed. The results showed that the LiDAR-based method for detecting 3D droplet distribution is feasible, fast, and environmentally friendly.

Keywords: LiDAR; droplet distribution; air-assisted spraying

check for updates

Citation: Wang, Z.; Zhang, Y.; Li, T.; Müller, J.; He, X. Visualization of Lidar-Based 3D Droplet Distribution Detection for Air-Assisted Spraying. *AgriEngineering* **2023**, *5*, 1136–1146. https://doi.org/10.3390/agriengineering5030072

Academic Editors: Marcello Biocca and Roberto Fanigliulo

Received: 30 April 2023
Revised: 13 June 2023
Accepted: 23 June 2023
Published: 3 July 2023

1. Introduction

Air-assisted spray technology is currently the most widely used pesticide application technology for plant protection in orchards. The basic principle is to deliver pesticide droplets to all parts of the canopy through strong airflow generated by a fan to achieve droplet deposition uniformly and improve pesticide efficacy [1]. The airflow of the fan not only increases the droplet deposition inside the canopy by opening the branches and leaves outside the canopy, but it also raises the adhesion of the pesticide droplets on the back of the leaves as a consequence of promoting the swing of the branches and leaves. In traditional orchard air-assisted sprayers and tower sprayers, adjustments to the wind field are typically made by modifying the length and angle of the deflector, as well as the mounting position. However, multi-channel orchard air-assisted sprayers provide more precise wind field adjustments by enabling the adjustment of the position and spray angle of each outlet. As the research in the field of air-assisted spraying has progressed, an increasing number of parameters have been found to be related to its effect, which has made it challenging to determine the optimal design of air-assisted spraying [2].

The conventional method for testing spray distribution in orchards was carried out based on the amount of droplet deposition, mainly using a vertical deposition distribution test bench [3]. However, this test method requires multiple tests at various distances to obtain a two-dimensional spatial distribution, and obtaining a three-dimensional spatial distribution is challenging. Due to the high cost of actual testing and the difficulty of reproducing the environment, much of the optimization work for the orchard sprayer was based on computational fluid dynamics (CFD). CFD-based technology can realize the simulation of two-phase flow to achieve the tracking of droplet particles and calculate the results of droplet spatial distribution [4,5], but there is still a discrepancy between CFD simulation outcomes and actual results [6].

Laser imaging is an imaging measurement technique that utilizes the reflection of particles in the air against an emitted light source. The most common laser imaging techniques include particle droplet image analysis (PDIA) and particle image velocimetry (PIV). PDIA uses a laser or monochromatic light source as a background to rapidly capture high-resolution images of droplets which are then identified and analyzed for the size, velocity, and direction of droplet motion using two consecutive teens [7,8]. However, the observable field of view of PDIA is too small to track a large range of droplets, and PDIA can only analyze droplets within the focal plane. On the other hand, PIV is a technique for the velocity analysis of droplet particles within the laser plane using high-speed photography, and this method has higher requirements for the camera as well as for the purity of the background during the computation [9,10].

LiDAR technology has been applied in agriculture as early as 1984 [11]. Since then, with the continuous advancement and optimization of LiDAR technology, its widespread adoption in agriculture has been increasingly observed [12–15]. Particularly in the 21st century, LiDAR technology has developed rapidly, and at the same time, with the gradual popularization of unmanned vehicles, the price of LiDAR has been gradually reduced, which makes it popular in agriculture, an extremely cost-conscious field. The multipurpose function of LiDAR is also gradually being explored, for example, Seol et al. achieved simultaneous target and drift detection using LiDAR mounted on a variable rate sprayer, showcasing its potential to optimize pesticide application and minimize environmental impact [16,17].

For drift testing, the use of lidar for spray drift testing was investigated in 1997, where the LiDAR was mounted on a horizontal and vertical rotating mechanism, respectively. This was achieved by scanning in a 2D plane using a pulsed LiDAR and using the LiDAR reflection signal intensity to obtain a 2D measurement of the plume [18]. However, the research on spray 3D distribution is currently focused more on drift measurement [19,20], with less emphasis on spray droplets. Li et al. used 3D LiDAR to achieve the droplet distribution measurement of a single nozzle and performed quantitative analysis [21], and Boqin Liu et al. used 3D LiDAR to perform droplet detection on the air-assisted unit and fitted the deposition based on the droplet density distribution [22]. However, there is still a lack of research on the processing of LiDAR visualization effects.

Therefore, this study proposed a visualization method based on LiDAR for droplet distribution measurement technology. By moving the machine perpendicular to the LiDAR scanning plane, the 2D LiDAR scanning data were reconstructed into a 3D spatial distribution. The three different installed nozzle positions and three different fan speeds were adjusted, the noise and background of the point cloud were removed, and the point cloud was aligned. The reconstructed 3D spatial distribution of the droplet cloud provides a fast, inexpensive, and visualized method for structural and application parameter settings for orchard air-assisted sprayers and similar machines.

2. Material and Method

2.1. Installation of LiDAR

The LiDAR (LMS111, Sick, Waldkirch, Germany) was mounted on a tripod at a height of 2.35 m above the ground, and the maximum detection of the LiDAR range is 20 m. The communication cable and power cable (12 V DC) were then connected to the device. The LiDAR was fixed on the tripod after rotating 90 degrees, and the vertical direction of the LiDAR was set as the X-axis parallel to the ground, and the vertical direction from the ground was set as the Z-axis. The travel route of the machine was planned 6 m away from the LiDAR, and the travel route was not less than 6 m. The travel direction is set as the Y-axis, perpendicular to the detection plane (XOZ), so as to establish a three-dimensional coordinate system. The LiDAR scanning frequency was set to 25 Hz, and the scanning angle was set at −45~225° with a 0.25° interval. The first echo signal was selected, and the fog filter was turned off.

2.2. Spraying System

The spraying system utilized in this experiment consisted of 3 main components: the air-assisted system, the liquid pump, and the nozzle. The air-assisted system includes a brushless motor (X5212S, Sunnysky, Columbus, OH, USA), a carbon fiber propeller (2055, T-Motor, Nanchang, China), an electronic speed controller (X-Rotor 80A-HV, Hobbywing, Shenzhen, China), and a motor signal generator (DC6. HJ Facalhobby, Shenzhen, China). The nozzle used was a TR8004 (Lechler, Metzingen, Germany) operating at a pressure of 0.3'MPa. To achieve 3D space detection with a 2D LiDAR, it is necessary to move the machine or LiDAR in a direction perpendicular to the scanning plane to obtain a similar result to CT tomography, which can then be reconstructed in 3D space. The whole system was powered by 220 V and the operator pushed the system along the predetermined route at a forward speed of 0.5 m/s, which is the Y direction in Figure 1. In this test, the fan was set at 1500 rpm, 2500 rpm, and 3500 rpm, and the axial wind speeds at 50 cm were 3 m/s, 7 m/s, and 10 m/s. The nozzle positions were provided with three mounting heights located at the edge of the fan: above the axis (H1), at the center of the axis (H2), and below the axis (H3), as shown in Figure 2. All spraying tests were conducted at night with an ambient wind speed of less than 0.5 m/s.

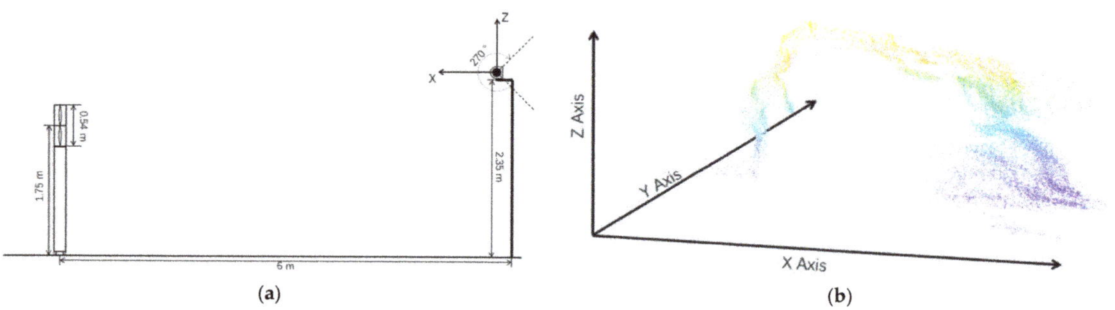

Figure 1. (**a**) LiDAR and sprayer setup layout (XOZ plane). Dotted line is the detected range of LiDAR (from 45~315° in XOZ plane). (**b**) Sprayer and droplet cloud in XYZ coordinate system. Color means the heights from yellow (highest) to blue (lowest).

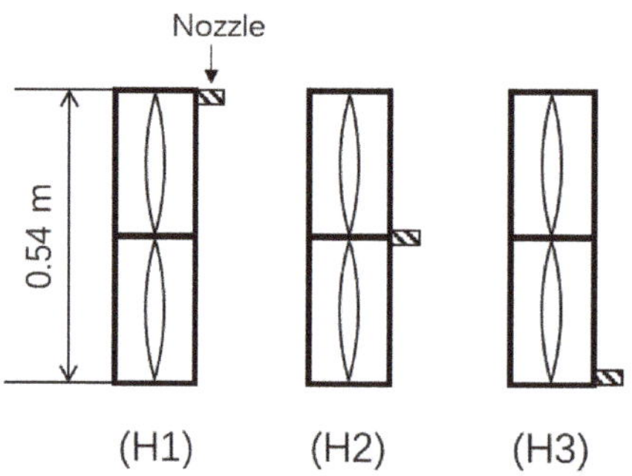

Figure 2. Nozzle setup layout. Above the axis (H1), center of the axis (H2), below the axis (H3).

2.3. Data Processing

2.3.1. Point Cloud Calculation from LiDAR

All data processing was performed in MATLAB 2022b (MathWorks, Natick, MA, USA) (CPU i7-8559U, RAM 32 GB). The XYZ coordinates of the point cloud were calculated based on the time and distance information obtained from the LiDAR data, as shown in Equations (1)–(3):

$$X_i = D_i \cdot \sin(\alpha_i) \quad (1)$$

$$Y_i = T_i \cdot v \quad (2)$$

$$Z_i = D_i \cdot \cos(\alpha_i) \quad (3)$$

where D_i and T_i are the original LiDAR data, which are the distance from point i to LIDAR, "mm"; scanning time, "ms". α_i is the scanning angle, which is known by checking the table in the manual, "°". v is the traveling speed of the machine, which is 0.5 m/s in this experiment.

Figure 3a,e shows the original point cloud data. The test region of interest (ROI) was selected, and the range was set to $-2000 < X < 8000$ mm. The preliminary ground fit was performed using the pcfitplane function [23]; the reference vector was set to [0, 0, 1], the and ROI range was set to min(Z) < Z < min(Z) + 200, and the ground splitting result is shown in Figure 3b,f. The ground obtained from the splitting process was compressed in the XOZ plane, and the ground slope was obtained by linear fitting and the ground tilt angle β. The point cloud was then rotated by β degrees around the y-axis using a transformation matrix [23]. The ground was re-fitted to the rotated point cloud using the pcfitplane function [24] with the same parameters as the first fit. The result of this second fit was shown in Figure 3c,g. The split ground information was then deleted and only the non-ground information was retained. The non-ground information was rotated by -β degrees around the y-axis using the transformation matrix to recover the droplet point cloud information, as shown in Figure 3d,h.

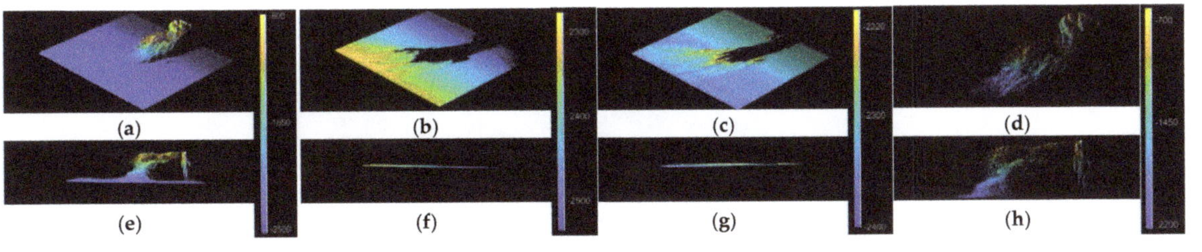

Figure 3. Point cloud results during the data processing, where (**a**–**d**) viewed at 45° from moving direction; (**e**–**h**) viewed at front direction (Y-axis). And (**a**,**e**) were original point cloud, (**b**,**f**) were after first ground fit, (**c**,**g**) were after second ground fit and rotating, and (**d**,**h**) were the non-ground results. Color means height (mm) from yellow (highest) to blue (lowest). Black is the background color to enhance the show, which has no meaning.

The pcsegdist function was used to remove the discrete noise points [25], with a minimum distance threshold of 500 mm. The part above X > 6000 mm is divided as the machine part, and the coordinate points [X_s, Y_s, Z_s] of the machine head are derived. The [X_s, Y_s, 0] point is set as the origin of the coordinate system by Equations (4)–(6).

$$X_i' = X_i - X_s \quad (4)$$

$$Y_i' = Y_i - Y_s \quad (5)$$

$$Z_i' = Z_i + Z_g \quad (6)$$

where the constant Z_g in the z-axis calculation is the height from the ground, which is 2350 mm in this experiment.

2.3.2. Triangulation for Point Cloud of Droplet

All point clouds were gridded at 100 mm intervals, and the duplicate data were removed. The detailed running code is shown in Code S1. The internal point cloud of the spray body was removed to retain only the surface points (Point Cloud S1). The surface points were triangulated using the alphaTriangulation function [26], and the triangulated results were subjected to Laplace smoothing [27].

3. Results

3.1. Segmentation Points Cloud

The point cloud splitting results are shown in Figure 4, with the ground rendered in white, the machine and operator in red, and the spray droplet portion represented by a gradient from yellow to blue, indicating different heights, with yellow being the highest and blue the lowest. The 3D display is shown in Video S1. It can be seen that the ground separation results were good, except for the R3200-H3 treatment, where some of the ground was not correctly identified. But almost all of the ground was identified accurately in the other treatments. For the machine and the operator, the results of the three iterations of the overlay are in the same position and the machine is correctly identified. The shape of the spray droplets was also more direct and clearer, which can effectively describe the spray droplet distribution for the different parameters.

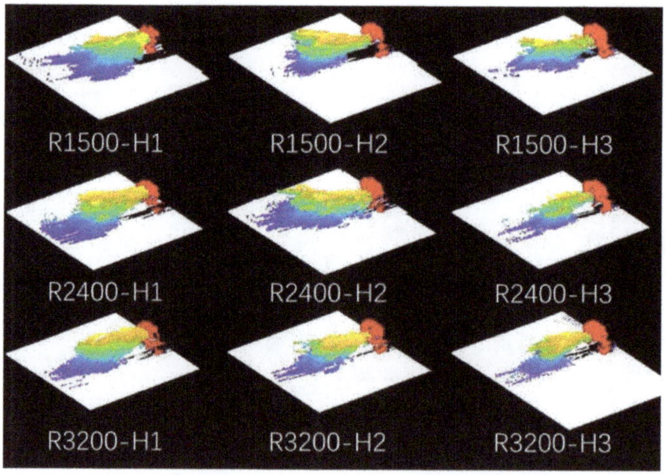

Figure 4. Integration result of 3 times repetition for all parameters, where white points mean the points from ground, red points mean the points from sprayer and operator, and the colorful points mean the points from droplet cloud. Where R1500 means the rotor speed is 1500 rpm; H1, H2, and H3 means different nozzle installed position at the top, middle, and bottom, respectively.

3.2. Repeatability of Multiple Spraying

The high overlap between the sprayer head and the operator in the three replicates provides evidence that using the head of the implement as the origin of the coordinate axis is a valid approach. By using the coordinate system with the sprayer head as the origin of the axis, the droplets from three different replicates exhibited good agreement (Figure 5, Video S2), indicating the reproducibility of this test method. Moreover, the shape of the spray from the same machine was approximately consistent, and the trajectories were similar in both spray distance (X direction) and travel direction (Y direction).

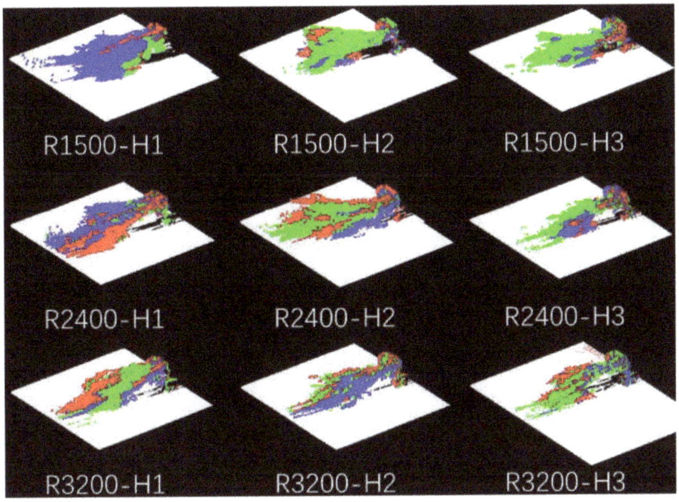

Figure 5. Integration result of 3 times repetition for all parameters, where white points mean the points from ground, red points mean the points from 1st repetition, green points mean the points from 2nd repetition, and the blue points mean the points from 3rd repetition.

3.3. Triangulation of Different Parameters

Figure 6 illuminates the results of the triangulation of different nozzle installation positions at varying fan speeds. The transparency setting of the droplet cloud indicates that the opaquer part is denser, whereas the less dense part is represented by the edge of the droplet cloud. In Figure 6a–c, the initial position of spray marked in a different color could tell the nozzle position correctly. The red color shows that the nozzle was installed in the upper part of the wind-fed spray and its droplet cloud starts higher than the green (H2, center) and blue (H3, bottom). And the rotation speed at a low speed (R1500) and high speed (R3200) did not significantly affect the wind-fed distance in the range of 1.8–4.5 m in the wind direction. However, in the top view of direction, the mounting height had little effect on the droplet cloud pattern under low-speed conditions (Figure 6d). Under medium- and high-speed conditions (Figure 6e,f), the higher the mounting height, the more backward the droplet cloud moved relative to the direction of travel. This result indicates that the transport of droplets was faster at higher mounting heights.

The effect of the fan speed at the different nozzle installation was shown in Figure 7. When the nozzle is installed on top (H1), increasing the fan speed from 1500 rpm to 2400 rpm can significantly increase the wind speed distance (Figure 7a) and reduce the backward distance of the droplets (Figure 7d), which can help the droplets reach the target faster and reduce potential drift. However, increasing the fan speed to 3200 rpm will not improve the range any further. When the nozzle is installed in the center (H2), the effect of changing the fan speed on droplet transport is not very significant (Figure 7b,e)). When the nozzle is installed at the bottom (H3), increasing the fan speed can slightly reduce the backward movement (Figure 7f).

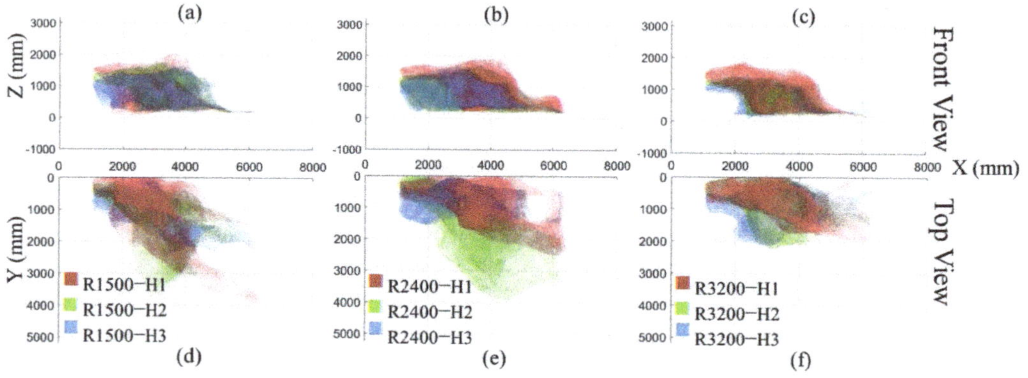

Figure 6. Front view (**a**–**c**) and top view (**d**–**f**) of three different nozzle position under different rotor speed, where R1500 means the rotor speed is 1500 rpm; H1, H2, and H3 means different nozzle installed position at the top, middle, and bottom, respectively.

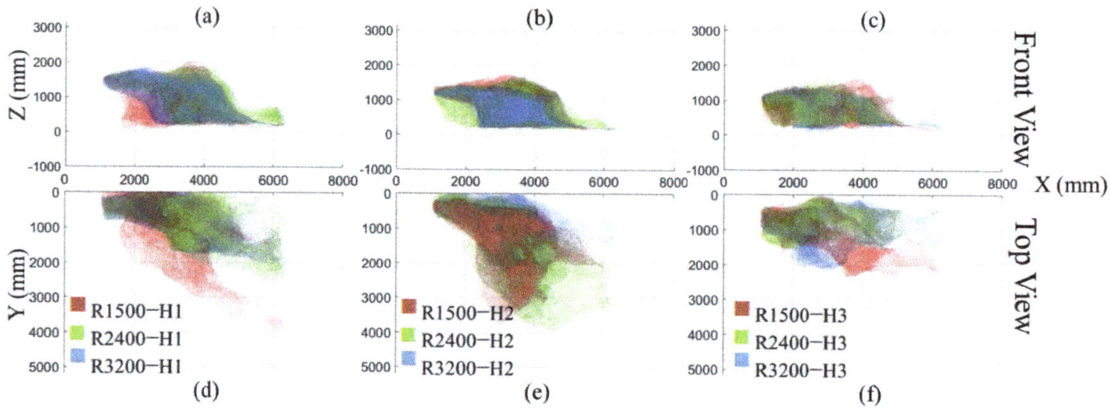

Figure 7. Front view (**a**–**c**) and top view (**d**–**f**) of three different rotor speed under different nozzle position, where R1500 means the rotor speed is 1500 rpm; H1, H2, and H3 means different nozzle installed position at the top, middle, and bottom, respectively.

4. Discussion

4.1. Droplet Distribution

Due to the lack of large-scale three-dimensional droplet distribution test methods, the development and parameter adjustment of orchard air-driven sprayers can only be carried out through droplet deposition tests. However, the test in the actual orchard needs to consume a lot of manpower and material costs. At the same time, due to the influence of the canopy growth cycle, a large number of test samples are required to improve the test accuracy. Therefore, a new method of droplet distribution visualization based on LiDAR-based droplet detection was proposed. Although quantitative analysis was not performed in this study, certain patterns could still be found by the variation in different parameters.

Normally, the nozzle was installed in the center of the air-assisted unit, but this is not always correct, i.e., the nozzle installed at the top (H1) can help the droplet delivery at the 2400 rpm fan speed (R2400), as shown in Figure 6b, as well as reduce the backward direction drift, as shown in Figure 6e. Three different fan speeds were set, but the delivery distance did not always increase with the higher fan speed (Figure 7), which is an important and valuable point that can save a lot of energy. This phenomenon can give manufacturers and users some ideas to improve the air-assisted sprayer.

This detection method is also applicable to the air delivery system of the axial fan, as well as the latest multi-unit air delivery spray system. It will greatly shorten the time for tool design parameters and spray parameter optimization. Compared with the traditional field test, this method is faster and does not require consumables. Also, the calculation method does not require a large computing power compared with CFD technology. In addition, with the recent popularization of unmanned vehicle technology, the price of LiDAR has gradually decreased. This research has good generalization and universality.

Current studies using LiDAR to detect droplets tend to focus on drift rather than droplets. The main methods used to quantify the distribution are droplet density [17,20,22,28] and emission intensity [29,30]. Since the spatial density of droplets cloud is much higher than that of drift, it is not easy for droplets to obscure droplets in the drift test, and the depth of the test field is sufficient to achieve droplet density detection over a larger area. However, for the detection of droplet distribution, the simultaneous detection of multiple locations may be required, and the algorithm will be fused to finally meet the all-around accurate detection of spatial distribution. While some LiDAR sensors can record multiple echo signals, potentially resolving mutual occlusion between droplets, this may also increase noise signals due to transmission and complex reflection within the droplet cloud, thereby exacerbating subsequent visualization processing difficulties.

Meanwhile, the reflection intensity of LiDAR is easily affected by the detection angle [21]. There has been even more in-depth research in the field of meteorology for the LiDAR detection of water droplet clouds [31]. The difference in the size and detection distance between these kind of raindrops and spray droplets is large, which can only provide partial theoretical support and cannot be directly applied to the field of agricultural spraying.

In conclusion, there are still many problems to be studied and worthy of research in the fog droplet distribution detection method of lidar in the future. Based on the visualization method of this study, the feasibility of this method has been preliminarily proven, and a series of application scenarios have been proposed and better prospects have been made.

4.2. LiDAR Droplet Detection Method

Because of the limitations of the current actual test methods, this test cannot be used as a confirmatory test, but the accuracy of this test can still vary from some aspects. For example, the spray start position can be clearly observed in Figures 6 and 7. The droplet cloud with the nozzle installed at the top (H1) started at about 1600 mm from the ground, as shown in Figure 7a; the nozzle installed at the center (H2) started at about 1350 mm, as shown in Figure 7b; and the nozzle installed at the bottom (H3) started at about 1100 mm, as shown in Figure 7c. This is consistent with the designed installation position of the nozzle (Figure 2), and the pattern is also consistent at the three different fan speeds, where the start positions almost completely overlap.

Whilst the size of the LiDAR (LMS111) spot diameter is approximately 1 mm, the size of agricultural spraying droplets is generally between 100 and 500 μm. When the spot is larger than the droplet particles, some of the laser light is reflected back to the LiDAR, while some of the laser light still passes through or bypasses the droplets. Therefore, the fog filter setting needs to be turned off in the LIDAR to prevent interference. For other LiDAR models, the fog droplets can be detected by selecting to accept the first reflection since they are the closest detected object. It is important to note that the LiDAR only records the first reflected light, resulting in a better detection effect for the surface of the droplet cloud closer to the LiDAR side compared to the inner and distal droplets cloud relative to the LiDAR. This effect can be observed in Figure 3b,c, where a shadow on the ground is visible after removing the droplet cloud. This shadow suggests that the LiDAR cannot fully penetrate the droplet cloud to detect objects on the other side, indicating a limitation in its detection capabilities. In light of the LiDAR's limited ability to detect objects beyond the surface of the droplet cloud near its side, there may be decreased detection efficacy for droplet cloud objects further from the LiDAR. As evidenced in Figure 7b, there is minimal variation in the 0–2 m range with an increasing fan speed, although this may be due to the reduced detectability of the LiDAR at this distance. Thus, careful consideration must be given to interpreting LiDAR data, particularly when evaluating objects located far from the LiDAR's position.

4.3. Accuracy of 3D Model Based on Evaluated Speed

According to Equation (2), the Y-axis data are calculated by speed and time, and the time recording is completed by the LiDAR itself, which has a high accuracy. The accuracy of the Y-axis is mainly affected by the forward speed; therefore, the forward speed used for scanning is crucial for achieving accurate results. The forward speed should be stable, and the operator must practice several times to achieve a stable forward speed. Higher forward speeds result in larger distances between scanning layers and lower spatial resolution. Conversely, lower forward speeds result in longer scanning times and higher static requirements for the scanned object. In addition, unstable environmental conditions may negatively impact the results obtained with lower forward speeds. To ensure accurate scanning of the droplet cloud, it is crucial to maintain its stability and stationary position during the scanning process. Therefore, the scanning should be conducted under stable environmental conditions, as was done in this study by selecting almost windless conditions. However, under windy conditions, it may be necessary to increase the number of repetitions and superimpose the results to obtain the final distribution, or to increase the step speed to reduce the scanning time. In this case, the LiDAR scanning frequency should be increased and the angular resolution reduced to compensate for the decrease in spatial resolution in the direction of the driving axis caused by increasing the step speed, i.e., sacrificing resolution in the scanning plane to improve resolution in the driving direction.

5. Conclusions

In this study, a LiDAR-based method was proposed for visualizing the 3D distribution of droplets in wind-driven spraying. The method involved scanning the droplet distribution at three different fan rotation speeds for three different nozzle installation positions using 2D LiDAR. The point cloud results were divided and noise was reduced for the open environment, and triangulation was performed to visualize the shape of the droplet cloud based on the fusion of multiple scans. The results showed that the proposed method was feasible, had good reproducibility, and the effect of different spray parameters on the droplet distribution could be analyzed based on the triangulation results. Although this study did not perform actual test verification, it still presents and discovers some regularities through experiments. Compared to the other measurement and simulation methods of the droplet distribution, the LiDAR results provided excellent visualization without requiring consumables or indicators, making it an environmentally friendly and fast detection method. However, more research is still needed to evaluate the accuracy of LiDAR for droplet distribution results, especially the experimental quantitative comparison of droplet density.

Supplementary Materials: The following supporting information can be downloaded at: https://www.mdpi.com/article/10.3390/agriengineering5030072/s1, Code S1: Triangulation processing; point cloud S1: Surface points cloud R1500 H1; Video S1: Figure 4; Video S2: Figure 5.

Author Contributions: Z.W.: conceptualization, investigation, methodology, validation, and writing—original draft. Y.Z.: conceptualization, investigation, and writing—review and editing. T.L.: conceptualization and investigation. J.M.: supervision and review. X.H.: supervision and funding acquisition. All authors have read and agreed to the published version of the manuscript.

Funding: This study was financially supported by the earmarked fund for CARS: CARS-28, the 2115 Talent Development Program of China Agricultural University, Sanya Institute of China Agricultural University Guiding Fund Project (Grant No. SYND-2021-06).

Data Availability Statement: To verify the possibility of using the data available in the Supplementary Materials or the images collected, please contact the first author.

Acknowledgments: The authors would like to thank Guoyi Ma for their support in the measurement.

Conflicts of Interest: The authors declare no conflict of interest.

References

1. Liu, L.; Liu, Y.; He, X.; Liu, W. Precision Variable-Rate Spraying Robot by Using Single 3D LIDAR in Orchards. *Agronomy* **2022**, *12*, 2509. [CrossRef]
2. Li, T.; Qi, P.; Wang, Z.; Xu, S.; Huang, Z.; Han, L.; He, X. Evaluation of the Effects of Airflow Distribution Patterns on Deposit Coverage and Spray Penetration in Multi-Unit Air-Assisted Sprayer. *Agronomy* **2022**, *12*, 944. [CrossRef]
3. Wegener, J.K.; von Hörsten, D.; Pelzer, T.; Osteroth, H.J. Basic Research into Different Parameters Influencing the Quality of Vertical Distribution by Crop Sprayers. *Landtechnik* **2016**, *71*, 4–13. [CrossRef]
4. Dekeyser, D.; Duga, A.T.; Verboven, P.; Endalew, A.M.; Hendrickx, N.; Nuyttens, D. Assessment of Orchard Sprayers Using Laboratory Experiments and Computational Fluid Dynamics Modelling. *Biosyst. Eng.* **2013**, *114*, 157–169. [CrossRef]
5. Duga, A.T.; Delele, M.A.; Ruysen, K.; Dekeyser, D.; Nuyttens, D.; Bylemans, D.; Nicolai, B.M.; Verboven, P. Development and Validation of a 3D CFD Model of Drift and Its Application to Air-Assisted Orchard Sprayers. *Biosyst. Eng.* **2017**, *154*, 62–75. [CrossRef]
6. Duga, A.T.; Dekeyser, D.; Ruysen, K.; Bylemans, D.; Nuyttens, D.; Nicolai, B.M.; Verboven, P. Numerical Analysis of the Effects of Wind and Sprayer Type on Spray Distribution in Different Orchard Training Systems. *Bound. Layer Meteorol.* **2015**, *157*, 517–535. [CrossRef]
7. Kashdan, J.T.; Shrimpton, J.S.; Whybrew, A. Two-Phase Flow Characterization by Automated Digital Image Analysis. Part 2: Application of PDIA for Sizing Sprays. *Part. Part. Syst. Charact.* **2004**, *21*, 15–23. [CrossRef]
8. Kashdan, J.T.; Shrimpton, J.S.; Whybrew, A. A Digital Image Analysis Technique for Quantitative Characterisation of High-Speed Sprays. *Opt. Lasers Eng.* **2007**, *45*, 106–115. [CrossRef]
9. De Cock, N.; Massinon, M.; Nuyttens, D.; Dekeyser, D.; Lebeau, F. Measurements of Reference ISO Nozzles by High-Speed Imaging. *Crop Prot.* **2016**, *89*, 105–115. [CrossRef]
10. Dorr, G.J.; Hewitt, A.J.; Adkins, S.W.; Hanan, J.; Zhang, H.; Noller, B. A Comparison of Initial Spray Characteristics Produced by Agricultural Nozzles. *Crop Prot.* **2013**, *53*, 109–117. [CrossRef]
11. Krabill, W.B.; Collins, J.G.; Link, L.E.; Swift, R.N.; Butler, M.L. Airborne Laser Topographic Mapping Results. *Photogramm. Eng. Remote Sens.* **1984**, *50*, 685–694.
12. Chen, Y.; Zhu, H.; Ozkan, H.E. Development of LIDAR-guided sprayer to synchronize spray outputs with canopy structures. In Proceedings of the American Society of Agricultural and Biological Engineers Annual International Meeting, Louisville, KY, USA, 7–10 August 2011; American Society of Agricultural and Biological Engineers: St. Joseph, MI, USA, 2011; Volume 1, pp. 11–24.
13. Gené-Mola, J.; Gregorio, E.; Guevara, J.; Auat, F.; Sanz-Cortiella, R.; Escolà, A.; Llorens, J.; Morros, J.-R.R.; Ruiz-Hidalgo, J.; Vilaplana, V.; et al. Fruit Detection in an Apple Orchard Using a Mobile Terrestrial Laser Scanner. *Biosyst. Eng.* **2019**, *187*, 171–184. [CrossRef]
14. Qiu, Q.; Sun, N.; Bai, H.; Wang, N.; Fan, Z.; Wang, Y.; Meng, Z.; Li, B.; Cong, Y. Field-Based High-Throughput Phenotyping for Maize Plant Using 3d LIDAR Point Cloud Generated with a "Phenomobile". *Front. Plant Sci.* **2019**, *10*, 554. [CrossRef] [PubMed]
15. Han, L.; Wang, S.; Wang, Z.; Jin, L.; He, X. Method of 3D Voxel Prescription Map Construction in Digital Orchard Management Based on LiDAR-RTK Boarded on a UGV. *Drones* **2023**, *7*, 242. [CrossRef]
16. Seol, J.; Kim, J.; Son, H. Il Field Evaluations of a Deep Learning-Based Intelligent Spraying Robot with Flow Control for Pear Orchards. *Precis. Agric.* **2022**, *23*, 712–732. [CrossRef]
17. Seol, J.; Kim, J.; Son, H. Il Spray Drift Segmentation for Intelligent Spraying System Using 3D Point Cloud Deep Learning Framework. *IEEE Access* **2022**, *10*, 77263–77271. [CrossRef]
18. Stoughton, T.E.; Miller, D.R.; Yang, X.; Ducharme, K.M. A Comparison of Spray Drift Predictions to Lidar Data. *Agric. For. Meteorol.* **1997**, *88*, 15–26. [CrossRef]
19. Gregorio, E.; Torrent, X.; Planas, S.; Rosell-Polo, J.R. Assessment of Spray Drift Potential Reduction for Hollow-Cone Nozzles: Part 2. LiDAR Technique. *Sci. Total Environ.* **2019**, *687*, 967–977. [CrossRef]
20. Li, L.; Zhang, R.; Chen, L.; Liu, B.; Zhang, L.; Tang, Q.; Ding, C.; Zhang, Z.; Hewitt, A.J. Spray Drift Evaluation with Point Clouds Data of 3D LiDAR as a Potential Alternative to the Sampling Method. *Front. Plant Sci.* **2022**, *13*, 939733. [CrossRef]
21. Tian, L.; Xiongkui, H.; Zhichong, W.; Huang, Z.; Han, L. Method for Measuring the 3D Spatial Distribution of Spray Volume Based on LIDAR. *Trans. Chin. Soc. Agric. Eng.* **2021**, *37*, 42–49. [CrossRef]
22. Liu, B.; Li, L.; Zhang, R.; Tang, Q.; Ding, C.; Xu, G.; Hewitt, A.J.; Chen, L. Analysis of the Spatial and Temporal Distribution of a Spray Cloud Using Commercial LiDAR. *Biosyst. Eng.* **2022**, *223*, 78–96. [CrossRef]
23. Mathworks Matrix Representation of Geometric Transformations. Available online: https://ww2.mathworks.cn/help/releases/R2022b/images/matrix-representation-of-geometric-transformations (accessed on 27 March 2023).
24. Torr, P.H.S.; Zisserman, A. MLESAC: A New Robust Estimator with Application to Estimating Image Geometry. *Comput. Vis. Image Underst.* **2000**, *78*, 138–156. [CrossRef]
25. Mathworks Pcsegdist. Available online: https://ww2.mathworks.cn/help/releases/R2022b/vision/ref/pcsegdist.htm (accessed on 27 March 2023).

26. Mathworks AlphaTriangulation. Available online: https://ww2.mathworks.cn/help/releases/R2022b/matlab/ref/alphashape.alphatriangulation.html (accessed on 27 March 2023).
27. Zhang, Y.; Hamza, A. Vertex-based anisotropic smoothing of 3D mesh data. In Proceedings of the 2006 Canadian Conference on Electrical and Computer Engineering, Ottawa, ON, Canada, 7–10 May 2006; pp. 202–205.
28. Gil, E.; Llorens, J.; Llop, J.; Fàbregas, X.; Gallart, M. Use of a Terrestrial LIDAR Sensor for Drift Detection in Vineyard Spraying. *Sensors* **2013**, *13*, 516–534. [CrossRef] [PubMed]
29. Gregorio, E.; Rosell-Polo, J.R.; Sanz, R.; Rocadenbosch, F.; Solanelles, F.; Garcerá, C.; Chueca, P.; Arnó, J.; del Moral, I.; Masip, J.; et al. LIDAR as an Alternative to Passive Collectors to Measure Pesticide Spray Drift. *Atmos. Environ.* **2014**, *82*, 83–93. [CrossRef]
30. Gregorio, E.; Rocadenbosch, F.; Sanz, R.; Rosell-Polo, J. Eye-Safe Lidar System for Pesticide Spray Drift Measurement. *Sensors* **2015**, *15*, 3650–3670. [CrossRef]
31. Sassen, K.; Zhao, H. Lidar Multiple Scattering in Water Droplet Clouds: Toward an Improved Treatment. *Opt. Rev.* **1995**, *2*, 394–400. [CrossRef]

AgriEngineering

Article

A Study of an Agricultural Indoor Robot for Harvesting Edible Bird Nests in Vietnam

Duc Anh Vu Trinh [1] and Nguyen Truong Thinh [2,*]

1 Department of Electrical Engineering, University of South Florida, Tampa, FL 33620, USA; ducanhtrinh@usf.edu
2 Institute of Intelligent and Interactive Technologies, University of Economics Ho Chi Minh City—UEH, Ho Chi Minh City 700000, Vietnam
* Correspondence: thinhnt@ueh.edu.vn

Abstract: This study demonstrates robot technology for harvesting edible bird's nests within swiftlet houses. A comprehensive manipulator's movement analysis of harvesting operation with a separating tool is provided for precisely collecting swiftlet nests. A robotic manipulator mounted on a mobile platform with a vision system is also analyzed and evaluated in this study. The actual harvesting or separating the swiftlet nests is performed with visual servo feedback. The manipulator performs the gross motions of separating tools and removing the nests under computer control with velocity and position feedback. The separating principle between the objective nest and wooden frame has been applied to a demonstration removal of nests using a four-degrees-of-freedom manipulator to perform the gross movements of tool. The actual separations using this system are accomplished as fast as the manipulator can be controlled to perform the necessary deceleration and topping at the end of separating. This is typically 2.0 s. This efficiency underscores the system's capability for swift and precise operation in harvesting an edible bird nest task.

Keywords: robot; harvest; edible bird's nest; swiftlet; Vietnam bird nest; robotic system; EBN

check for updates

Citation: Trinh, D.A.V.; Thinh, N.T. A Study of an Agricultural Indoor Robot for Harvesting Edible Bird Nests in Vietnam. *AgriEngineering* **2024**, *6*, 113–134. https://doi.org/10.3390/agriengineering6010008

Academic Editor: Simone Pascuzzi

Received: 27 November 2023
Revised: 27 December 2023
Accepted: 4 January 2024
Published: 12 January 2024

1. Introduction

In the past, robots were mainly for increasing productivity and expanding production capacity under fast growing necessaries. But, in recent years, we are facing the necessity of developing a new concept for the introduction of robots into harvesting swiftlet nests. We understand that one of the new goals of the robots from social issues is to promote automated robots instead of labor increase. On the other hand, humans make mistakes when harvesting swiftlet nests. In this case, the decision to select a particular operation to be automated is made only by considering the technological possibility and economical desirability. Today, manual harvesting of swiftlet's nests in the house causes many obstacles to the survival and reproduction of swiftlets. Moreover, manual harvesting completely depends on the experience of the harvester. The environment in the swiftlet farming area is quite toxic affecting human health, so the trends of automation in monitoring and harvesting of nests in the house is being researched and implemented. The edible bird nests are harvested from natural caves, and they are taken increasingly from purpose-built brick houses, which are designed and structured to attract swiftlets.

There are many species of avian that use its saliva grande as a component in nest buildings, but the swiftlet might be the extreme case where its nest is made entirely of the saliva. Out of numerous swiftlet species, namely the edible-nest swiftlet (*Aerodramus fuciphagus*), the black-nest swiftlet (*Aerodramus maximus*), the Atiu swiftlet (*Aerodramus sawtelli*), and others, only some of them whose nests are edible belong to the family *Aerodramus fuciphagus*. They frequently breed in the ceiling of caves or the edge of cliffs where they are not exposed to direct sunlight, but the majority of edible birds' nests are now collected in artificial bird's nest houses for higher productivity and lower hazard. The white translucent house's nest is made of glutinous saliva solution that securely clings to vertical walls and is hardened over time. Nutritional studies [1,2] have shown EBN is mainly made of protein and carbohydrates with very low input of fat. Economically, EBN can fetch more than USD 2500 per kilogram, mostly produced in Southeast Asia, and the global bird's nest industry is estimated to be worth USD 5 billion, mainly consumed in countries like China, Hong Kong, and Taiwan. The garnering of EBN in bird's nest houses has yielded elevated income due to the nest's expensive tag and reduced life-threatening risks related to cliff-climbing for farmers recently. However, manually harvesting the EBN, like most of the agricultural field, is repetitive, time-dependent, and labor-intensive, and can even have negative effects in maintaining the house bird's population owing to the swiftlet's dislikes of unfamiliar human odors. These difficulties can be resolved by an EBN harvesting robot since as proved by many real-world examples, utilizing a farming robot is faster, more accurate, and more efficient because of its inexhaustive and controllable characteristics.

Currently, in Southeast Asia, domestic swiftlet farming is being developed. Houses or factories for raising swiftlets are designed and built depending on investment costs and farming area. Like Indonesia [3], most of the swiftlet houses built in Vietnam are brick houses built with wooden trusses for birds to nest (Figure 1). The swiftlet house is built with a diverse scale with different number of floors from one to four floors. The minimum size of the nesting rooms is usually 8.0 × 16 m or 40 m. The height of each floor is usually 2 m, sometimes up to 2.5 to 3.0 m high. The formation of the nest begins with nesting and egg laying. The mother bird builds the nest by dripping saliva. When the nest is formed, the mother begins to lay eggs and then incubate. The chicks are incubated and hatched in the nest and cared for by their parents until they are ready to fly.

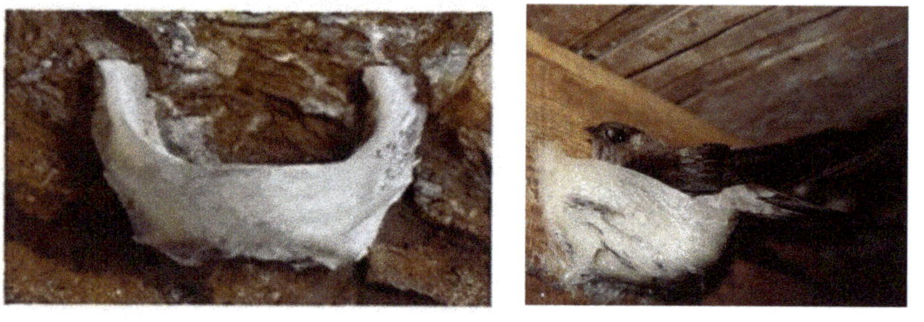

Figure 1. Edible swiftlet nests in nature and in-house.

Since each agricultural product is harvested in a different environment, researchers have been developing solutions aimed at specific commodities. This is reasonable because each species has particular strength, size, and shape, which can make the method for one kind damaging for another. Although for many crops and fruits, agricultural robots are extensively studied and catered for, there is sparse research about harvesting robots that aim at EBN. Hence, it is of profound importance to develop a robotic system designed to automate the bird's nest harvesting process of in-house environments [4].

A robot consists of different sub-systems such as the mobile platform, manipulator, and end-effectors which are designed to match the task requirement and environment constraints. Each subsystem is reviewed from various literature, and this study's automated system is presented thereafter. Smart agriculture is the use of smart technology applied to agriculture with the aim of saving labor and improving product quality and productivity. The caring for and raising of swiftlets for nesting is an agricultural profession developed in Vietnam. Therefore, the care, monitoring and harvesting should use AI with Machine learning to ensure sustainable development in agriculture.

A study [5] makes a review of the design of manipulators used in harvesting robots, which shows that electric servo and stepper motors are mostly utilized when the weight load is light, whereas the custom electric actuator or hydraulic motor is used in the opposite situation. Additionally, [5] also lists three requirements to consider when designing the manipulator, which are adaptable enough for the working environment, vigorous enough for the required forces, and sustainable enough against the environment's condition. To correctly control the manipulator, kinematics and inverse kinematics models need to be calculated, so [6] utilizes the Denavit–Hartenberg model for the path planning of cucumber harvesting. The end effector is the interaction part between the product and the robot, so its design needs to be delicate enough to not impair the product but powerful enough to extract the product from the main body. According to [7], the fruit harvesting method commonly utilizes one made out of product grasping, stem grasping, rotation mechanism, cutting devices, and pressure suction as its main development. Grasping, suction, and rotation alone cannot be applied to EBN harvesting since the approach will easily break the nest due to the nest's strong adhesion to the wooden cell of EBN house. The cutting mechanism, however, is harder to implement in fruit harvesting because good stem detection is required. Refs. [8,9] develop a thermal blades solution by creating a voltage between electrodes in which both uphold the gripper structures to stabilize the product before the cutting process. Meanwhile, reference [10] chooses a high-speed mechanical rolling blade to cut off the pumpkin's stem. Four-wheel vehicles are dominant in developing a harvesting robot [11] because of their light weight and flexibility for dampish conditions. Likewise, track vehicles are less sensitive to the ground environment, so it can work well in swampy and wet ground quality [12]. Due to more pressure on the ground, the crawler-type wheel has an advantage when working on rough terrain compared to that of the four-wheel [13]. Six-wheel steering is also used when high maneuverability is demanded on unpredictable terrain. Additionally, ensuring that all wheels sustain contact with the ground in any circumstances is a crucial design requirement. Ref. [14] designs a flexible platform frame that is slightly bendable when contacting with inclined ground. Many systems for fruit harvesting in a greenhouse use a rail mobile platform to travel due to the organized rows of crops, such as in [12]; so the path is always determined and slipping is minimized compared to the wheel-type system. If the wheel-type or crawl-type wheel is used, path planning incorporated with landmark detection needs to be executed. As such, [13] with crawl-type wheel uses the GNSS and LiDAR sensors to determine the location of ground turn and orchards row-turn respectively, and the robot's path is tracked continuously. Path planning for a robot system working on farmland is developed in [14]. Many systems in greenhouses use rail mobile platforms to travel due to the organized rows of crops, such as in [15]. Therefore, the path is always determined and slipping is minimized compared to wheel-type system. If the wheel-type or crawl-type wheel is used, path planning incorporated with landmark detection needs to be executed. As such, [16] with crawl-type wheels uses the GNSS and LiDAR sensors to determine the location of ground turn and orchards row-turn respectively, and the robot's path is tracked continuously. Due to the flourishing of machine vision, a large number of harvesting solutions use vision-based methods. When combining color and geometric features of the target, fruits like tomato, apple, citrus, and lichi get correctly detected with a precision over 90% as mentioned in [17] which provide a review of object recognition approach. Nevertheless, camera-based still faces difficulties in ambient light and various illumination conditions, so LiDAR-based is frequently selected as an alternative. Ref. [18] applies cartographer algorithm in ROS for local SLAM and loop-back detection to LiDAR's point cloud, and Publish Point function in ROS combined with Dijkstra algorithm is applied for global and local path planning. Path planning constrained by different limits for a robot system working on farmland is developed in [19].

The swiftlet bird builds its nest from saliva strands, but only some of the swiftlet species' nests are proved to be edible and harvested, which majorly comprises glycoprotein to which galactose, galactosamine, and other carbohydrates attach [20]. Moreover, the nest cement has enough adhesive properties when dried up to hold against its weight and the bird's egg. Specifically, the nest's adhesive strength overall can load a vertical force of around 0.465 N considering the heaviest case scenario of two birds', two eggs', and the nest's weights. Paper [21] calculates the maximum stress in this scenario to be 0.56 MPa at the rim where the birds stand. However, there is a limited source for the adhesiveness of the nest to the wooden frame partly because the sample is normally analyzed in the laboratory. Therefore, the actual force to separate the nest in the harvesting site is not regenerated. When each component is brought together, a control unit fuses the data and decides the reaction of those peripherals. Ref. [22] provides the general control system for apple harvesting robot, but the graph can be applied to most harvesting robots.

2. Swiftlet-Nest Harvesting Robot

This study introduces the design and planning of movement paths based on reinforcement deep learning to address the significant challenge of harvesting swiftlet nests in a combined environment of structured and unstructured spaces, ensuring that the robot can harvest targeted bird nests without colliding or interacting with neighboring nests. The recurrent neural network is applied to recollect and find the past state observed by the computer vision system and determine depth according to the gradient algorithm to yield the appropriate movement. The task of this study is to plan a collision-free path, which consists of a sequence of regression reference points, into the mixed environment after the position of the interested object is determined.

Figure 2 depicts the structure of the designed bird's nest harvesting robot. It consists of a moving platform, vision system, and manipulator, which are the main parts of swiftlet-nest harvesting robotic system. The vision system and manipulator are both integrated and firmly fixed on the moving platform. The mobile platform is made of an aluminum platform with three wheels including two active wheels and a caster wheel moving according to differential drive. The moving platform, manipulator, and vision system are controlled by a central controller unit (CPU) that communicates with the computer via a universal asynchronous receiver–transceiver (UART). The moving platform carries the robot arm and vision system to the desired position under the control of the CPU. The robotic structure is designed to be suitable for the size of the birdhouse space as well as the workspace of the manipulator. The nests are mostly located on the walls of the swiftlet-house, so harvesting nests is a sequence of the tool's movements to the position beneath the nest, followed by a lateral movement along the wall to perform the harvest. Thus, the required manipulator should have four degrees of freedom: three translational movements in cartesian space and one rotation about the vertical axis to be able to apply the tool's force to the wall. To perform the harvesting function, a device, similar to a scraper, is used to separate the nest from the wall, which is considered as the end-effector of the manipulator. The bird's nest separating tool resembles a shovel, while the cup mimics a human bird's nest harvester. To expand the workspace of the manipulator, the tool is placed on a moving platform, effectively extending the operational space reach within the swiftlet-house. To perform the bird's nest recognition, positioning, and navigating the robot, the camera system can translate in the vertical axis and is also mounted on the moving platform. Therefore, the field of view of the robot can be greatly expanded. Swiftlet's nest has long been a culinary delight in many countries and tagged with a high price due to its supposed health benefits. Traditionally, they are harvested from natural cliffs, which is extremely dangerous and requires safety gears to collect high-built nests up in the cave's ceiling. Therefore, man-made houses are constructed to duplicate the natural living condition of the swiftlet so that the bird is attracted to nest in this guarded space. The swiftlet first builds an anchor from its saliva gland that is hardened and adhered to the high wooden house's frame when exposed to the air. Thereafter, it weaves the nest around the spot to form a semi-cup shape that firmly sticks to the frame along the nest's boundary [23]. To harvest bird nests, the collectors need to have a specialized knowledge of the suitable time and nest's maturity level. Because of the nest's persistent adhesiveness, the collectors usually use a thin scraping tool to gently detach the nest from the frame by delicately pushing and picking between the nest and its gluing surface. As such, the same method with a non-abrasive and fine tool is applied as the robot's end effector to undisturbed applied force from the bottom of the nest. When there is sufficient force, the nest-frame adhesive bond is breakable, and the nest can be collected.

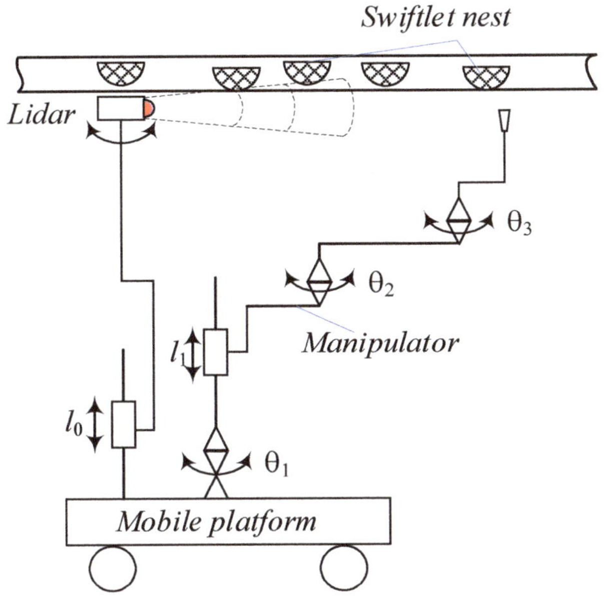

Figure 2. Modeling of manipulator.

Due to the asynchronous formation of nests, not all of them can be harvested, which constitutes a significant challenge in this study. The selection of the candidate bird's nest to harvest is based on its size and formation time to ensure quality and precise operation. The weight of the bird's nest is very light, and it is very easy to break due to strong impact force. Therefore, careful attention, flexibility, and meticulousness must be given to ensure precise and accurate harvesting operations when employing a robot for this task. To harvest the bird's nest, the end-effector, or a tool as simple as a blade, can separate the bird's nest from the wooden frame of the house, safely assumed to be flat surfaces.

The separation process of harvested swiftlet nests from the surface of frames is completed by using the tool attaching manipulator. In order to analyze the process in detail, its two-dimensional modeling is presented in Figure 3. In this idealized modeling the separating tool moves bottom to top along the swiftlet nest at a constant velocity V. The nest is removed from the frame by a tool based on the shearing of the nest's adhesive layer continuously along the shear plane. The forces and power involved in separating operations is significantly important since power requirements must be known to enable the calculation of the manipulator's dynamics with adequate power. Equally crucial, the separating forces are required without excessive distortion so that the nest's shape is intact. The forces acting on the tool are shown in Figure 3. The cutting force F_c acts in the direction of the removing speed V and supplies the energy required for the harvesting process. The thrust force F_t acts in the direction normal to the separating velocity, that is, perpendicular to the frame. These two forces produce the resultant force R. The resultant force can be resolved into two components on the tool face: a friction force F along the tool-nest interface and a normal force N perpendicular to it. Another way is to resolve into the shear force F_s along the shear plane and a normal shear force $F_{s,\,n}$ perpendicular to it. This paper assumes that the width of the adhesive layer before and after the shearing process is almost identical. Therefore, if the rake angle, α, is between the tool's surface and the perpendicular direction of the frame, then the shear angle, β, is calculated by (1), and the shear force is determined by (2).

$$\tan^{-1}\left(\frac{\cos(\alpha)}{1-\sin(\alpha)}\right) \tag{1}$$

$$F_c\cos(\beta) - F_t\sin(\beta) \tag{2}$$

During the harvesting operation of the manipulator, the separating tool attaching to the manipulator can reach the swiftlet nests in its workspace. In this study, a manipulator named modified SCARA with four degrees of freedom was used as a harvesting manipulator. A modified SCARA manipulator is generally represented in the form of an open loop kinematics linkage. It includes four degrees of freedom presenting a multivariable mechanical system. The manipulator should copy the movements of the human arm harvesting the edible swiftlet's bird in the house. The number of degrees of freedom of the arm is in the range of 6 ≥ n ≥ 3. The movement of two human arms holding the swiftlet separating tool for harvesting is quite flexible and has high DoFs. The number of DoFs is modeled, and each joint must have a reference system of axes that includes the z and x axes. There are several concepts, but the modified SCARA is selected with three rotating joints, and a translation one. We can model the coordinate system of the manipulator (Figure 4) according to the rotation using the anatomy and representation of the coordinate system and the rotation joints in the human body. Each joint is given a set of coordinate axes.

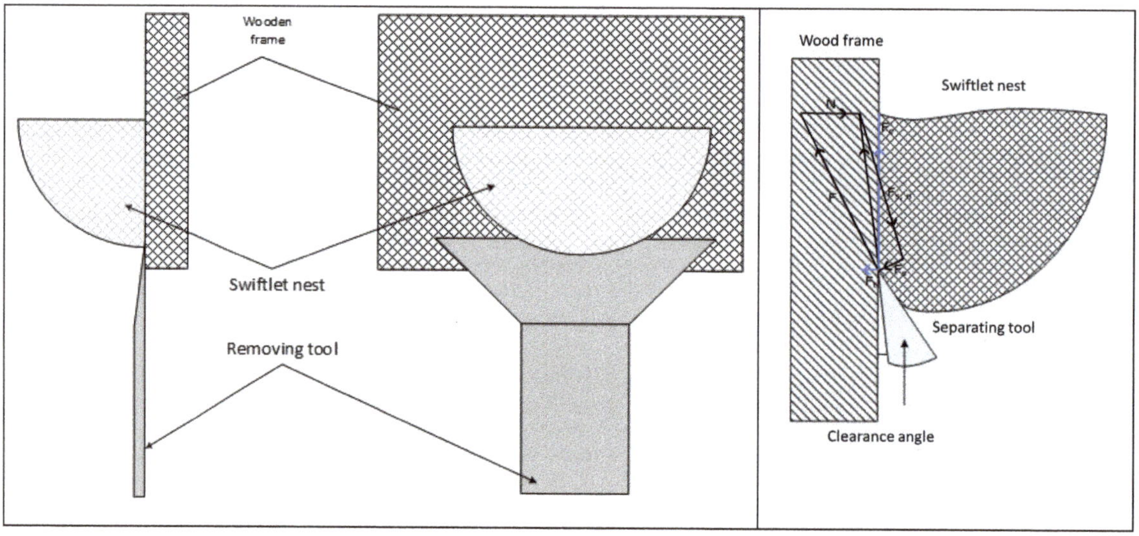

Figure 3. Force analysis on 2D representation of harvesting process.

Before a dynamic system is constructed, it is essential to carry out the analysis and synthesis on the mathematical model of the system. Like the concept presented in the previous section, it described an end-effector moving along a vertical wooden wall. The task space coordinate system is given in the u and v directions. In this case, the tool is moved along the surface in the v direction and normal force is applied to the surface along the u direction. The vector of task space is x, the correlation defined the forward kinematics is defined as follows:

$$\mathbf{x} = \begin{bmatrix} x \\ y \\ z \\ \beta \end{bmatrix} = \varphi(q_i) \quad \boldsymbol{i = 1 \ldots 4;} \tag{3}$$

where $\varphi : \mathbb{R}^3 \to \mathbb{R}^3$.

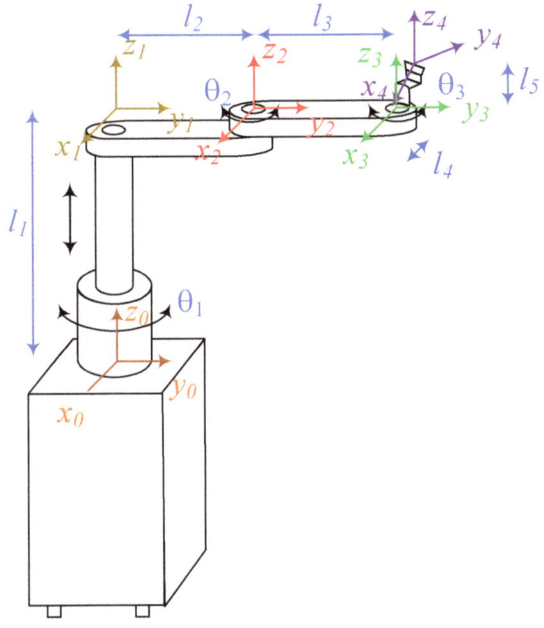

Figure 4. Kinematic model with assigned frames based on D–H rules.

In this study, the coordinate systems for the joints and links of the 4-degrees-of-freedom manipulator follow the Denavit–Hartenberg (D–H) rule. Since the four axes of joints are parallel, the axes are assigned as shown in Figure 4. During the bird's nest harvesting operation of the robotic manipulator, the separating tool of manipulator can reach the nest locations so that it can be harvested in the workspace and perform harvesting operation in-sequence. With the configuration of the modified SCARA manipulator of 4 degrees of freedom, it can perform translational movements along the x, y, z axes and rotation around the z axis. The manipulator is generally presented in the form of an open loop kinematics. Theoretically, SCARA would be required to separate an object attaching the other one with a desired orientation to find the position of the end-effector based on the rotation angles of the manipulator's joints and is a forward kinematics problem. The parameters of the D–H table are based on the poses of the coordinate attached to the joints. The coordinate transformation between two consecutive frames i and frame $i + 1$ is obtained as follows.

According to the parameters of the manipulator based on the D-H method, the forward kinematics equation of the manipulator is formed by multiplying the matrix of transformation matrices. The result of the forward kinematics equation is as below.

$$ {}^{n}_{n+1}T = \begin{bmatrix} \sin(\theta_{n+1}) & -\sin(\theta_{n+1})\cos(\alpha_{n+1}) & \sin(\theta_{n+1})\sin(\alpha_{n+1}) & a_{n+1}\cos(\theta_{n+1}) \\ \sin(\theta_{n+1}) & \cos(\theta_{n+1})\cos(\alpha_{n+1}) & -\cos(\theta_{n+1})\sin(\alpha_{n+1}) & a_{n+1}\sin(\theta_{n+1}) \\ 0 & \sin(\alpha_{n+1}) & \cos(\alpha_{n+1}) & d_{n+1} \\ 0 & 0 & 0 & 1 \end{bmatrix} \quad (4) $$

$$ {}^{0}T_4 = {}^{0}T_1{}^{1}T_2{}^{2}T_3{}^{3}T_4 $$
$$ = \begin{bmatrix} \cos(\theta_1+\theta_2+\theta_3) & -\sin(\theta_1+\theta_2+\theta_3) & 0 & -l_2\sin\theta_1 - l_3\sin(\theta_1+\theta_2) \\ \sin(\theta_1+\theta_2+\theta_3) & \cos(\theta_1+\theta_2+\theta_3) & 0 & l_2\cos\theta_1 + l_3\cos(\theta_1+\theta_2) \\ 0 & 0 & 1 & l_1 \\ 0 & 0 & 0 & 1 \end{bmatrix} \quad (5) $$

When the robot arm is in operation, the separating tool must be correctly controlled with the required speed of displacement and vertical upward movement. The impact forces and velocity of the tool are carefully calculated to ensure the quality of the harvested bird's nest. After multiplying all the matrices together and simplifying the equation, the positions of the tool can be determined as follows:

$$P_x = l_2 \cos\theta_1 + l_3 \cos(\theta_1 + \theta_2) + l_3 \cos(\theta_1 + \theta_2 + \theta_3) \tag{6}$$

$$P_y = l_2 \sin\theta_1 + l_3 \sin(\theta_1 + \theta_2) + l_3 \sin(\theta_1 + \theta_2 + \theta_3) \tag{7}$$

$$P_z = l_1 + l_5 \tag{8}$$

$$\beta = \theta_1 + \theta_2 + \theta_3 \tag{9}$$

For the task of the nest's separation from wooden frames by the manipulator, the position controller gives a responding performance because the task only requires the robot to follow a predetermined required trajectory. However, the manipulator contacts with the environment, so there are the interaction forces between the separating tool mounted on the end-effector and the environment, so the interaction force and position of the separation tool need to be controlled. To separate the nests for harvesting, the position of the tool should be controlled. That is, insufficient press can result in adhesion that is not easily separable, while excessive pressure can lead to the nest's breakage. Where $\phi(q)$ is forward kinematics, the derivative of x is defined.

$$\dot{\mathbf{x}} = J(\mathbf{q})\dot{\mathbf{q}} \tag{10}$$

The relationship between the speed of the manipulator's end-effector and velocities of angular and linear joints can expressed with the following equation by Jacobian matrix.

$$\dot{\mathbf{x}} = \begin{bmatrix} \mathbf{v} \\ \tilde{\omega} \end{bmatrix} = J(q)\dot{q} \tag{11}$$

In the study, the manipulator which uses the angular and linear joints can be shown.

$$\begin{bmatrix} \mathbf{v} \\ \tilde{\omega} \end{bmatrix} = \begin{bmatrix} z_i \times {}^i p_n^0 \\ z_i \end{bmatrix} \dot{q} \tag{12}$$

$$J_i = \begin{bmatrix} z_i \times {}^i p_n^0 \\ z_i \end{bmatrix} \tag{13}$$

where J_i is the ith column of Jacobian matrix, p is the position vector of the origin in the coordinate system of the manipulator relative to the coordinate system $\{i\}$ in the base coordinate system. The z_i denotes the unit vector of the z-axis of the coordinate system $\{i\}$ in the base coordinate system. The dynamic model of the manipulator with 4 degrees of freedom, rigid, with no friction at the joints is shown below.

$$M(\mathbf{q})\ddot{\mathbf{q}} + C(\mathbf{q}, \dot{\mathbf{q}})\dot{\mathbf{q}} + \mathbf{g}(\mathbf{q}) = \tau \tag{14}$$

The state vector $\begin{bmatrix} \mathbf{q}^T & \dot{\mathbf{q}}^T \end{bmatrix}$ is modified as follows:

$$\frac{d}{dt}\begin{bmatrix} \mathbf{q} \\ \dot{\mathbf{q}} \end{bmatrix} = \begin{bmatrix} \dot{\mathbf{q}} \\ M(\mathbf{q})^{-1}[\tau(t)] - C(\mathbf{q}, \dot{\mathbf{q}})\dot{\mathbf{q}} - \mathbf{g}(\mathbf{q}) \end{bmatrix} \tag{15}$$

The Jacobian matrix in $n \times n$ task space is defined as follows:

$$J(\mathbf{q}) = \begin{bmatrix} I & 0 \\ 0 & T \end{bmatrix} \frac{\partial \varphi(\mathbf{q})}{\partial \mathbf{q}} \tag{16}$$

where, I is the identity matrix and T is transformation matrix.

In addition, we can determine the Jacobian matrix with four variables as joints. From the positions and orientation obtained, the relation between the velocities is shown in below.

$$\begin{bmatrix} \dot{x} \\ \dot{y} \\ \dot{z} \\ \dot{\beta} \end{bmatrix} = \begin{bmatrix} \frac{\partial x}{\partial \theta_1} & \frac{\partial x}{\partial l_1} & \frac{\partial x}{\partial \theta_2} & \frac{\partial x}{\partial \theta_3} \\ \frac{\partial y}{\partial \theta_1} & \frac{\partial y}{\partial l_1} & \frac{\partial y}{\partial \theta_2} & \frac{\partial y}{\partial \theta_3} \\ \frac{\partial z}{\partial \theta_1} & \frac{\partial z}{\partial l_1} & \frac{\partial z}{\partial \theta_2} & \frac{\partial z}{\partial \theta_3} \\ \frac{\partial \beta}{\partial \theta_1} & \frac{\partial \beta}{\partial l_1} & \frac{\partial \beta}{\partial \theta_2} & \frac{\partial \beta}{\partial \theta_3} \end{bmatrix} \begin{bmatrix} \dot{\theta}_1 \\ \dot{l}_1 \\ \dot{\theta}_2 \\ \dot{\theta}_3 \end{bmatrix} = J(\mathbf{q}) \begin{bmatrix} \dot{\theta}_1 \\ \dot{l}_1 \\ \dot{\theta}_2 \\ \dot{\theta}_3 \end{bmatrix} = J(\mathbf{q}) \tag{17}$$

And the Jacobian matrix is determined as follows:

$$J(\mathbf{q}) = \begin{bmatrix} \begin{matrix} -l_2 \sin\theta_1 - l_2 \sin(\theta_1 + \theta_2) \\ -l_2 \sin(\theta_1 + \theta_2 + \theta_3) \end{matrix} & 0 & \begin{matrix} -l_2 \sin(\theta_1 + \theta_2) \\ -l_2 \sin(\theta_1 + \theta_2 + \theta_3) \end{matrix} & -l_2 \sin(\theta_1 + \theta_2 + \theta_3) \\ \begin{matrix} l_2 \cos\theta_1 + l_2 \cos(\theta_1 + \theta_2) \\ +l_2 \cos(\theta_1 + \theta_2 + \theta_3) \end{matrix} & 0 & \begin{matrix} l_2 \cos(\theta_1 + \theta_2) \\ +l_2 \cos(\theta_1 + \theta_2 + \theta_3) \end{matrix} & l_2 \cos(\theta_1 + \theta_2 + \theta_3) \\ 0 & 1 & 0 & 0 \\ 1 & 0 & 1 & 1 \end{bmatrix} \tag{18}$$

where $J(\mathbf{q}) = \frac{\partial \varphi(\mathbf{q})}{\partial \mathbf{q}} \in \mathbb{R}^{4\times4}$ is called the analytical Jacobian matrix, and the relation of acceleration also called the differential kinematics is as follows:

$$\begin{bmatrix} \ddot{x} \\ \ddot{y} \\ \ddot{z} \\ \ddot{\beta} \end{bmatrix} = \left[\frac{d}{dt} J(\mathbf{q}) \right] \begin{bmatrix} \dot{\theta}_1 \\ \dot{l}_1 \\ \dot{\theta}_2 \\ \dot{\theta}_3 \end{bmatrix} + J(\mathbf{q}) \begin{bmatrix} \ddot{\theta}_1 \\ \ddot{l}_1 \\ \ddot{\theta}_1 \\ \ddot{\theta}_1 \end{bmatrix} \tag{19}$$

The inverse kinematics model of the harvesting manipulator allows for obtaining the joint positions $\mathbf{q}$ in terms of the positions and orientation of tool attaching on the end-effector in the reference frame. The inverse kinematics is below:

$$\begin{bmatrix} \theta_1 \\ l_1 \\ \theta_2 \\ \theta_3 \end{bmatrix} = \varphi^{-1}(x, y, z, \beta) \tag{20}$$

where $\varphi^{-1} : \mathbf{q} \to \mathbb{R}^4$ and $\mathbf{q} \subseteq \mathbb{R}^4$.

The derivation of the inverse kinematics model is very complex and contrasts with the forward kinematics. The angular and linear accelerations and velocity of joints are determined according to recursive relationships. For each joint, inertial forces and the moments are calculated, and equations of kinematics are established. The configuration of the separating tool as the end-effector is described by a position and an orientation.

In order to reduce the impact force on the manipulator during operation as well as improve movement's stability, it is required that the trajectory of the end-effector and active joints of the manipulator have to be continuous and smooth. Therefore it should not have any sudden change in velocity and acceleration. When the separating tool harvests the nests, the path of the end-effector is optimally designed based on the information provided by the LiDAR system. The LiDAR collects the point clouds and then calibrates the position and orientation to give a suitable map. Information about the nesting parameters is located in the three-dimensional coordinates of swiftlet nests in workspace. After completing harvesting for each nest, the robot system returns to its initial position. The operation of harvesting the swiftlet nest is divided into three stages as follows. First, the separating tool is moved to the target (below the nest). Secondly, it is moved along the vertical axis from bottom to top. Finally, the tool is moved back to its original position. At each stage in the harvesting process, path planning is performed based on quintic polynomial interpolation. To ensure that harvesting does not affect the quality of the bird's nest, the speed of the tool must be controlled. The transport processes play an essential role at automated swiftlet nest harvesting, especially in moving the manipulator to suitable position determined by LiDAR's data, when the mobile platform is to be transported after finishing one operation to the next to be performed. The mobile platform moves along the stationary wall in the house facilities' handling of individual harvesting based on mapping from LiDAR. Works provided with a permanent internal transportation system are consequently forced to produce their products in long series and are so unable to adapt flexibly, without a cost increase, to the requirements of customers. The mobile robot has a virtual tracking system which conveys information as robot deviation from the direction and position given by the map to the control system.

In this study, a wheel mobile robot on the ground is used, its wheels move by only rotation and no slip in driving and lateral directions. There are two concepts for designing the mobile platform as differential and tracking drive. The differential drive is simpler than the other one. The differential drive is always in contact with the ground by three contact points. The drive has the advantages over the tracking drive due to larger contact surface between its wheels and ground, assuming no wheel slipping when changing direction. Therefore, a tracking drive is often used for rough ground with big obstacles which the robot cannot overcome. We need to integrate the robot's movements over time to determine its position accurately. Additionally, this research section considers the dynamics of the mobile robot system to address the challenges of controlling the robot's motion in an environment that involves impact forces. Even though wheel slip is a factor that reduces the accuracy of evaluating and estimating the robot's motion, when traveling at low speed, the effect of wheel slip relative to the road surface can be ignored. Wheeled robots move along the x-axis and the robot cannot move along the y-axis, and the position of the robot can be determined as the midpoint between the two active wheels. When the robot is moving, we will assume rolling without slipping on flat ground at low speed (5 km/h); therefore, the force of inertia in all directions is less than the force of friction. Under the differential drive system, the motion of the vehicle is determined by the control of wheel velocities. In order for both drive wheels to roll properly, the robot must rotate around a point situated on the common axis of the two wheels. By adjusting the relative velocity between the two wheels, the location of this rotation point can be changed, allowing for different trajectories of the vehicle. From the forward kinematics, it is possible to calculate the pose of the robot at any time t based on the control of the velocity of the left wheel and the right wheel. In general, the differential drive robot's ability for moving in a specific direction $\theta(t)$ at a certain velocity is governed by the following equations.

$$x(t) = \frac{1}{2}\int_0^t (v_r(t) + v_l(t))\cos(\theta(t))dt \tag{21}$$

$$y(t) = \frac{1}{2}\int_0^t (v_r(t) + v_l(t))\sin(\theta(t))dt \tag{22}$$

$$\theta(t) = \frac{1}{l}\int_0^t (v_r(t) - v_l(t))dt \tag{23}$$

Therefore, the angular and linear velocity is calculated in the equation below:

$$\begin{bmatrix} \mathbf{V} \\ \omega \end{bmatrix} = \begin{bmatrix} \cos(\theta) - \left(\frac{P_y}{P_x}\right)\sin(\theta) & \sin(\theta) + \left(\frac{P_y}{P_x}\right)\cos(\theta) \\ -\left(\frac{1}{P_x}\right)sin(\theta) & \left(\frac{1}{P_x}\right)\cos(\theta) \end{bmatrix} \begin{bmatrix} K_{P_x}\left(x_{ref} - x_P\right) \\ K_{P_y}\left(y_{ref} - y_P\right) \end{bmatrix} \tag{24}$$

where linear and angular velocity are **V**, ω of the robot.

The dynamics of the robot are determined by the non-holonomic constraint. In this study, a differential drive mobile robot is the approached object. The robot's center of gravity is placed in the midpoint between two wheels. Each wheel is controlled independently by the motor and the castor wheel has the ability to keep the robot's balance while it is moving on the flat floor. Our robot is shown in (Figure 5) with $2b$ as the distance between two wheels, R is the radius of each wheel, θ is the angular between x-axis and robot orientation. Assume that the robot does not slide on any axis. From this assumption, the robot is described below.

$$\dot{x}_C \cos\theta - \dot{y}_C \sin\theta = 0 \quad (25)$$

$$\dot{x}_C \cos\theta + \dot{y}_C \sin\theta = \frac{r}{2}(\dot{\varphi}_r + \dot{\varphi}_l) \quad (26)$$

The equation of dynamics is written by the following:

$$M(q)\ddot{q} + C(q,\dot{q}) + G(q) + \tau_d = B(q)\tau + A^T(q)\lambda \quad (27)$$

where $M(q)$ is the inertial matrix, A^T is the Jacobi matrix, $B(q)$ is the transpose matrix, τ is the moment, λ is the force vector, $C(q,\dot{q})$ is the matrix containing radial and rotational components, $G(q)$ is the gravity matrix, τ_d is the disturbance elements, and q is the status vector.

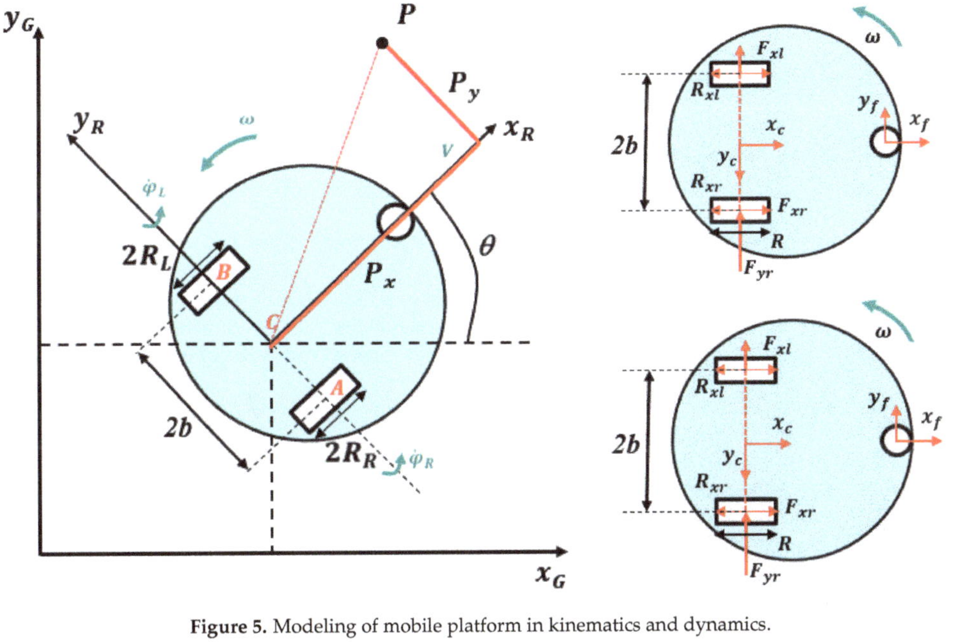

Figure 5. Modeling of mobile platform in kinematics and dynamics.

Scanning System Based on LiDAR

Dissimilar to the camera, LiDAR is not susceptible to dim or dark light conditions in the bird's nest house, so it is employed in this study. A 3D map shown in Figure 6 is created from mounding distinct height levels of 2D layers that are generated from LiDAR's point cloud for the vision capabilities of the robot. Since the data can be obstructed by both the environment's condition and its calibration, which introduce noise, outlier points are removed by counting the number of the surrounding points and comparing it to a limit. As each 2D layer of the wooden frame is obtained, the line segment algorithm is applied to detect the frame portion between two nests, and the nest intersection is the complement of the frame line. Consequently, a 3D segment of the bird's nest is acquired, and the nest's position can be interpolated for harvesting process.

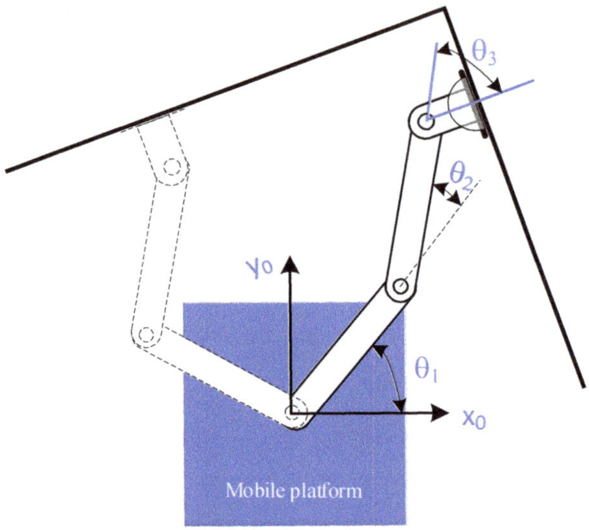

Figure 6. Schematic diagram of harvesting operation.

Line segmenting from the point cloud algorithm has been researched for many years with different approaches, so utilizing which method is a matter of choice regarding complexity and accuracy. Reference [22] reviews multiple research studies about line extraction from 2D sets of points and compares those algorithms on the same scanned LiDAR data. The incremental algorithm idea, as mentioned in [23], is used in this study to extract points that are classified as belonging to a line. The pseudo code for the algorithm is as follows:

- An initial *1...n* number of points is fit to a line, added to current set, and the RMSE is calculated by (8) and compared whether it is less than a user-specified threshold. In this study, the threshold is chosen based on the standard deviation of the statistical error of the LiDAR points.

$$\sqrt{\sum (f(x) - Y)^2} \quad (28)$$

- If 1. is true, the next point is added to the current set, and the change in RMSE is recalculated to test whether it belongs to the current set.
- If next point cannot be added, either the length of current set needs to be larger than a threshold to be considered a line segment, or the current set is discarded.
- If 1. is false, points *2...n + 1* are used instead of *1...n* and step 1. is repeated.

3. Controlling the Robot for Harvesting

3.1. End-Effector Control to Get the Bird Nest

Presently, the best available method of performing most precision swiftlet nest operations, and in some cases the only available method, is manual harvesting labor. This paper describes a method based on engineering compliance which permits the manipulator to perform swiftlet nest harvesting operations quickly, simply, and economically. The actual harvesting or separating the swiftlet nests is performed with visual servo feedback. The manipulator performs the gross motions of separating and removing the nests with velocity and position feedback under computer control (Figure 6). The principle of separation between the object and wooden frames has been implemented in a demonstration of nest removal. This was achieved using a four-degrees-of-freedom manipulator to execute the broader movements of the tool The actual separations using this system are accomplished as fast as the manipulator can be controlled to perform the necessary deceleration and topping at the end of separating, which is typically 2.0 s.

To reduce the workload and improve work efficiency, the bird's nest harvesting robot was developed with mobility and flexibility. By using computer vision and a manipulator placed on a mobile platform for bird nest identification and harvesting, laboratory experiments have verified the features of the robot bird's nest. However, this study still entails certain limitations in the performance of the robot, requiring further research. Firstly, the robot works in a confined environment whose power relies on lithium-ion batteries, which affects the time of use as well as the power of the actuators. During the harvesting operation, the computer vision and lidar system remains inactive, rendering it impossible to determine whether the harvest process has been completed and whether the quality of the harvested swiftlet nest is affected. In addition, the uneven surface of the swiftlet house floor due to the solidified bird waste makes it difficult to steadily roll.

3.2. Control Position and Velocity of Manipulator

The simplest way to indicate the movement of a manipulator is a method called point to point. However, in controlling the harvesting of bird's nests process, the separation device is required to go beneath the surface of the nest to adhere to the wooden wall, which requires a path in state space and its time parameterization to achieve. In the study, the movement path is defined as a straight line in state space connecting the initial point and the ending one to generate the required harvesting path. After obtaining the moving path, the inverse kinematics problem is applied to solve for the parametrical trajectories over time for control in the joint coordinates. Functional capabilities of a robot are broadly enlarged when a robot is supplied with sensing properties, but at the same time, supplying a robot with such properties demands the use of an artificial intelligence (AI) system. According to the algorithmic control concept, basing itself on the levels of normal dynamics and perturbed regimes, the manipulator control divides itself, due to its specificity into a normal dynamics level; level of perturbed regimes. The driving system of a manipulator is generally an active multi-link kinematic chain, which performs its action through a separating tool that interacts directly with the swiftlet nests to be manipulated. Any AI system of a sensible robot would include a whole set of standard "behavior" which is suitable to a wide range of situations and which, so far as it is possible. The tool directly interacts with the swiftlet nests to be manipulated. The operation of separating objects is a motional task combining joint working motions. Let us consider a case when the motion of the tool is determined by a sequence of Aj (j = 0, 1, . . ., m) as Appendix A. The coordinates θ_i describing the state of the kinetic chain, are connected to the coordinates x_i of the tool by below equations.

$$F_{\varsigma}(\theta_1, \ldots, \theta_m) \quad \xi = 1, \ldots, 4 \tag{29}$$

To separate the swiftlet nests from wooden frames, the movement of the tool is vertical on wooden frames. The line of tool movements has now been reduced to finding the sequence of configurations. Those configurations correspond to the positions of the tool in points A_j. Let us introduce the notion Ω to be called volume of motion.

$$\Omega = \sum_{i=1}^{n} a_i \int_{t_0}^{t_m} |\dot{\varphi}_i(t)| dt \tag{30}$$

where a_i is the coefficient of the ith coordinate. For pointwise description of tool motion, the functional may be expressed in the following form:

$$\Omega = \sum_{i=1}^{n} a_i \left(\sum_{j=0}^{m-1} |\varphi_i(t_{j+1}) - \varphi_i(t_j)| \right) \tag{31}$$

where t_j is the moment of the time when the tool coincidence with the point A_j. We admit that within the intervals (t_j, t_{j+1}) the generalized coordinates φ_i of the system vary monotonously.

3.3. Path-Planning of Mobile Robot

During the working process, the robot moves according to a predetermined trajectory combined with information from LiDAR to generate a suitable separating trajectory. The control system is driven by two individual electric motors controlled by signals issued by a map as well as tracking sensors. The mobile robot is propelled by these motors driving two wheels by differential drive. Servo motors applied as actuators in the moving platform ensure the control system can be used as driving. The mobile robot in autonomous driving can perform traveling along a predetermined path and stopping at calculated positions where the manipulator can reach to harvest the swiftlet nests. Moreover, the robot is capable of making stops at the charge docking station and negotiating crossover. It begins with the detection of growing nest sites by LiDAR and decides whether to harvest based on technical requirements of their sizes and dimensions. The positions and sizes of the nests are also collected through the time determined by the LiDAR in global coordinates, which is necessary for the evaluation of the harvest time as well as the use of the blockchain later. In monitoring operation, the mobile platform only carries LiDAR scanning in a zig-zag trajectory to generate the point cloud with an up and down movement along the x-axis. This is only to collect information to monitor the formation as well as change of performances of nests in the house. In addition, algorithms to identify and determine the parameters of the nest such as the positions and sizes are developed. After identifying the nests that can be harvested, a trajectory for harvesting the nests is given and decided by the quality evaluation. The harvesting is always reviewed and verified to determine the ID of nests. In cases of the nests located near each other, the harvesting process is affected due to the large size of the separation tool, which fore the operating trajectory to be adjusted accordingly. The trajectory of the bird's nest harvest can be described by three points as the starting point to contact the bird's nest at the lower position close to the nest, an intermediate point, and the ending point of the trajectory. Finally, the controlled path is interpolated to ensure smooth movement. The block diagram of the localization system of nest harvesting robotic system is presented in Figure 7. The separating tool is attached on the end-effector perpendicular to the floor. The initial position of the tool is A_i below the nest; then, it moves straight up to B_i, as follows.

$$A(x_1, y_1, z_1) \rightarrow B(x_2, y_2, z_2)(z_2 > z_1) \tag{32}$$

The system of parametric equations is a line, which represents the moving path of the tool as follows.

$$\begin{cases} X(t) = x_1 + (x_2 - x_1)\frac{t}{T} \\ Y(t) = y_1 + (y_2 - y_1)\frac{t}{T} \\ Z(t) = z_1 + (z_2 - z_1)\frac{t}{T} \end{cases} \tag{33}$$

From the above equation, angle θ_1 is also calculated as follows.

$$\theta_1 = a\tan 2\begin{pmatrix} (X(t) - l_3\cos\beta)(l_2 + l_3\cos\theta_2) + l_3\sin\theta_2(Y(t) - l_3\sin\beta), \\ (Y(t) - l_3\sin\beta)(l_2 + l_3\cos\theta_2) - l_3\sin\theta_2(X(t) - l_3\cos\beta) \end{pmatrix} \tag{34}$$

Therefore, angle θ_3 is also calculated depending on the determined orientation angle β and the two previously solved angles are θ_1 and θ_2. And the displacement length of the prismatic joint l_1.

$$l_1 = Z(t) - l_5 \tag{35}$$

The feature is extracted from point clouds with c being the coefficient representing the smoothness of the local planar surface [24]:

$$c = \frac{1}{|S| \cdot \|{}^{B}p_i\|} \left\| \sum_{j \in S, j \neq i} \left({}^{B}p_i - {}^{B}p_j \right) \right\| \quad (36)$$

where ${}^{B}p_i$ is an ith scanning point cloud $\{B\}$ point, and S is a set of consecutive points of ${}^{B}p_i$ in same chanel. The scanned point cloud from LiDAR includes concentric circles with a variety of heights and the center of LiDAR. The purpose of the LiDAR's scanning position with variable height is to build a 3D point cloud of the scanned environment and get the size and identification of the indexed bird's nests. The data of the point cloud is analyzed and calculated to determine the features of the surrounding nest and obstacles. The transformation of scans is also known as LiDAR odometry to solve nonlinear interations by minimizing the distance. At each scanning robot position, LiDAR creates a suitable 3D point cloud map, which is based on the previously presented algorithm to identify the frames as well as the nests. Calibration of data from LiDAR is performed and adjusted accordingly. When the system determines that the robot environment is operating with the prior map, the scan is also continuously updated to match with the prior map. The robot's position prediction lies on the manipulator's and platform's sensors to determine the robot's current position and correct data from obtained point cloud.

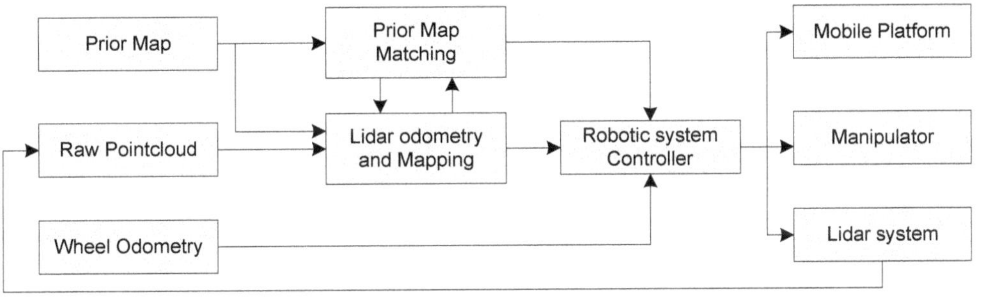

Figure 7. Block diagram of the localization system of nest harvesting robotic system.

The matching generated maps and prior maps provide pose estimation by the corresponding points between current point cloud and the reference point cloud. The transformations are to minimize the distance between pairs of correspondence points.

$$\mathbf{T} = \underset{\mathbf{T}}{\operatorname{argmin}} \sum_{i}^{N} d_i^{(\mathbf{T})^T} \left(C_i^B + \mathbf{T} C_i^A \mathbf{T} \right)^{-1} d_i^{(\mathbf{T})} \quad (37)$$

where $d_i^{(\mathbf{T})} = \|\mathbf{T}.a_i - b_i\|^2$ is distance between correspondence points; T is the predicted transformation between point clouds; a_i and b_i are corresponding points belonging to two point cloud sets; and C_i^A, C_i^B are covariance matrices for point cloud sets. The prior map contains information about the initial environment and LiDAR generates point cloud maps and keyframes of instantaneous data from which to calibrate the response to the position and orientation relative to the prior map. After comparing the LiDAR map and prior map, the localization and mapping data is determined for each robot's instantaneous position. The data is transferred to the robot controller for calculating the inverses of both the mobile platform and the manipulator for the joint variables to harvest the bird's nest. The operation is a nonlinear multivariable control.

4. Experiments and Discussions

4.1. Experiment Design

Experiments involved the use of robots to harvest edible bird nests and evaluating the performances of robots to perform the task (Figure 8). The robot is used to efficiently locate and extract bird nests from hard-to-reach places such as the bird's house ceiling and the ceiling's corner where harvesters face difficulties and need to use ladders. By using robots, it minimizes the disturbance caused to birds and their nesting environments, reducing stress and potential harm to the living environment. Moreover, future experiments can involve the collection of data about bird nesting habits, population trends, and habitat conditions by equipping robots with cameras, sensors, or other instruments to gather this information. The robot monitors swiftlets and their nests, helping to protect, manage, and identify their populations. We validated the design of a harvesting robotic system that runs LiDAR-based positioning and a mapping algorithm in a laboratory simulated swiftlet's house. According to the requirements based on the characteristics of the birdhouse, the operating space of the manipulator must be within 0.5 m × 0.5 m, which means the largest theoretical harvest area has a radius of 0.5 m. The LiDAR was used to acquire point clouds in the testbed, scanning 3D point cloud data automatically. The controllers of the harvesting manipulator and mobile platform are also designed and developed in accordance with the predetermined requirements, computing interpolation points and processes information. The snippet of the result for a layer after going through the line segment is shown in (Figure 9, dots). Afterward, a number of points among extracted points is randomly chosen to regress a line until a set loop threshold because a line is normally segmented into different pieces due to statistical deviation. Finally, the line with the lowest root mean square error is selected as the frame line. Nest is subtracted from the datapoint by taking the points exceeding the distance-to-line restriction and applying another noise filter. The nest point is also fitted with regression for enhanced accuracy of size estimation. After the points corresponding to the nest are extracted, the DBSCAN algorithm is applied to cluster the nest so that each nest layer is correctly grouped to form a completely differentiated 3D nest. Moreover, an aligned axis bounding box is also found around each nest for future application of machine learning to find the weight of the nest, which will be presented in another study.

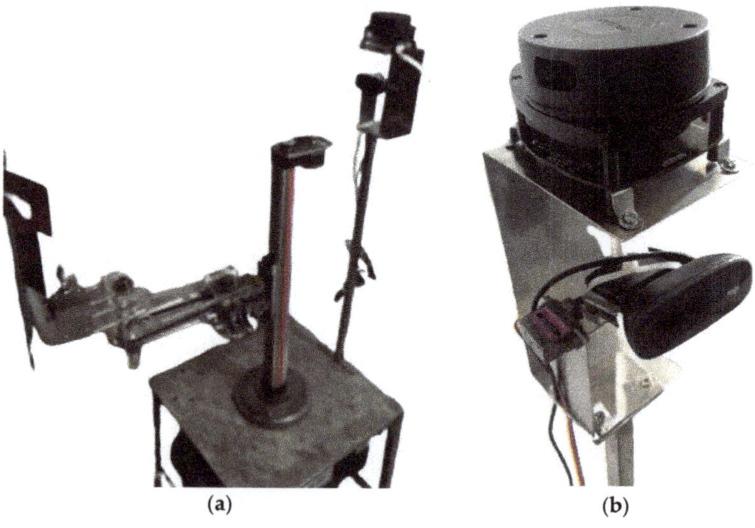

Figure 8. Robotic system is developed in this study (**a**), lidar and camera system for monitor and check (**b**).

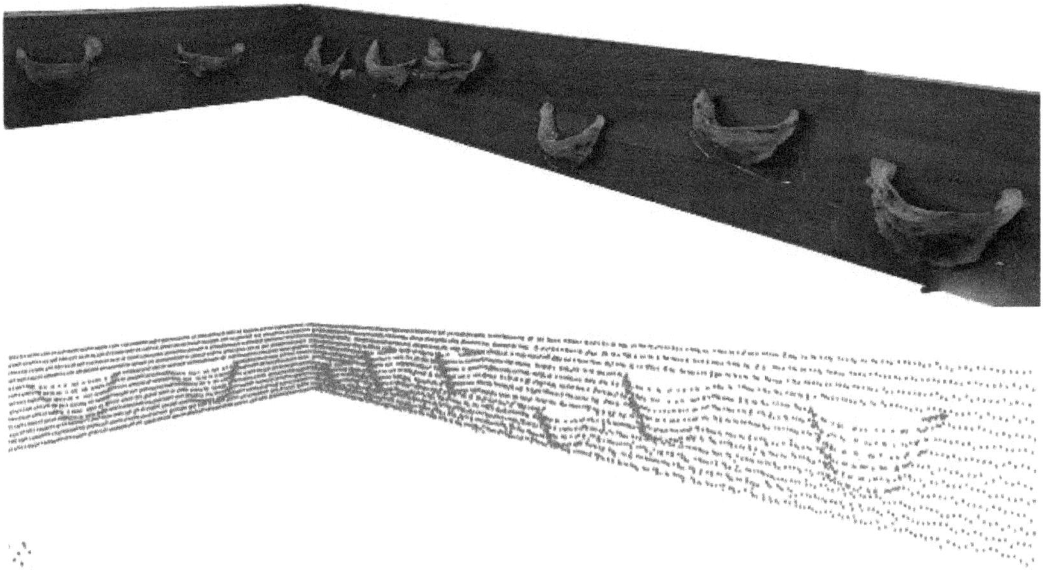

Figure 9. Edible Bird Nest transformed from real world to LiDAR's point cloud.

Line segmenting from point cloud algorithm has been researched for many years with different approaches, so utilizing which method is a matter of choice regarding complexity and accuracy. Study [24] reviews multiple research studies about line extraction from 2D sets of points and compares those algorithms on the same scanned LiDAR data. The incremental algorithm idea, as mentioned in [25], is used in this study to extract points that are classified as belonging to a line. Afterward, a number of points among extracted points are randomly chosen to regress a line until a set loop threshold because a line is normally segmented into different pieces due to statistical deviation. Finally, the line with the lowest root mean square error is selected as the frame line. The nest is subtracted from the datapoint by taking the points exceeding the distance-to-line restriction and applying another noise filter. The nest point is also fitted with regression for enhanced accuracy of size estimation.

4.2. *Evaluation*

The evaluation of harvested bird nest quality represents a critical aspect of indoor management and research, so a harvesting robotic system was prototyped according to the designed and simulated results. This process involves the thorough examination and assessment of the nests collected by the robotic system. The robot assesses the structural integrity of the harvested nest as the nest's size, shape, and overall condition, so any damages or irregularities are noted. The robot may also use both LiDAR and the camera to detect the presence of eggs, chicks, or even adult birds within the nest because they are crucial aspects of evaluation. The robotic system records the indoor location of the harvested nest, which is essential for understanding the distribution of bird nests and their preferences within a given habitat. The evaluation may also include an assessment of the environmental impact caused by nest harvesting or eggs' viability. Researchers analyze whether the robot's actions have disturbed the local ecosystem and take appropriate steps to minimize any negative effects. All collected data, including LiDAR-based positioning and mapping and other relevant information, is logged and stored for analysis. With the aim of testing in experiments the performance of the controller is shown on the manipulator harvesting swiftlet nests. After detecting the nests that need to be harvested, the location of each nest is determined through map matching, which detects features in maps at various times. The LiDAR system is capable of locating bird nests with acceptable accuracy. A lightweight manipulator was used to harvest swiftlet nests on a wooden frame to verify the performance of the manipulator in a bird's house in Khanh Hoa province. The robot was asked to harvest six nests in a space of simulated swiftlet's house as shown in Figure 10. Before harvesting, the LiDAR is used to identify the position of the center of the swiftlet nests.

Figure 10. The experimental setup used to evaluate robot's performance.

The harvesting duration of the manipulator is defined as the time from the beginning of detecting the nest to the finish of harvesting the nest based on its position and size. During the harvesting of the nests, the manipulator is started from the initial position to the end of Region of Interest (ROI), which is defined by LiDAR scanning. After individually harvesting the nests is completed, the manipulator returns to the initial position. The joint angular rotations are measured by encoders whose values are calculated the kinematics as position, orientation of the separating tool. The harvested bird nests are indexed and allowed to be harvested from the system. The harvesting process is to separate the nest stuck on the wall using a separating tool with distributed force while moving the tool on each surface at a changing speed in a trapezoidal shape to separate the nest without affecting the quality of the nest. To achieve this requirement, it is necessary to determine the exact center position of each bird's nest. During the mobile platform's movements, prediction and updating algorithms are used. It depends greatly on the accuracy of the LiDAR processing system. The separating controller determines when and which harvests accurately tracked the nests.

Experiments were implemented mainly in the laboratory where an indoor wooden frame system is built to simulate reality based on the technical required size. Due to the prohibitive cost of procuring bird's nest and high likelihood of non-consumable impurities after experiments render the nest non-reusable, the experimental bird's nests are made from plastic. First, the bird's nests are scanned with a 3D scanner, then the plastic nests are 3D printed and glued to a wooden frame as in a real environment for experimentation.

The locations of bird's nests are also randomly distributed like reality. Attaching the bird's nests to a wooden frame makes the adhesive force of this replicated nest greater than the authentic bird's nest because the natural bird nests' adhesive is saliva. When harvesting bird's nests, it is necessary to determine which nests would be harvested or not. Harvesting nests is determined by monitoring history and information obtained from LiDAR. The monitoring system knows which nests do not contain eggs or chicks. Monitoring the features of the nest helps to better evaluate the quality of the nest. This information is learned as well as proved appropriate harvesting methods later.

4.3. Discussion

When implementing the harvest process of an imitated bird nest, its quality is ensured characteristics such as nest's integrity, so the position of the separating tool needs to downsize its error to 1–2 mm. The robot moves to the appropriate position determined from saved LiDAR point cloud data. The nest's position from the LiDAR scan is used in the inverse kinematics problem calculation of the robot's moving platform and manipulator so that the separating tool is precisely moved to the beneath of the nest. Eventually, a vertical trajectory is created to separate the nest from the wooden frame, which has the dimension of 1.2×0.5 m. The robot is instructed to harvest from 10 to 20 nests unevenly attached to the frame. Accordingly, the experiment scores 100% accuracy of the nest's position, with 1.2 mm error in the x and y axis and 0.7 mm error in z axis. The difference in axis's errors is due to the fact that the maneuver in the z axis is affected only by translational joints, whereas the x and y axis comprise three joints, leading to higher inaccuracy compared to the other axis.

Our robot design is suitable for the harvesting edible swiftlet nest task in a closed environment such as bird house. However, the height information of the bird's nest can be falsely calculated if there are obstacles blocking the robot such as rocks or solidified bird's droppings. Furthermore, the robot only determines its position and direction mainly based on encoders affixed to two wheels, along with integrated sensors such as vison and LiDAR systems, relying solely on the local coordinate without input from the global coordinate system. The manipulator operates in a workspace within a radius of 0.5 m and only harvests swiftlet bird's nests in this confined area. The process of identifying bird nests from points in the cloud ignores location errors stemming from LiDAR information.

5. Conclusions

A flexible and autonomous bird's nest harvesting robot used in the laboratory has been designed, developed, and experimented with. The structure of the robotic hardware and control system is described, and tests are carried out to re-validate the functions and operations of the robot in the harvesting simulation environment. The results obtained from the experiment designed above show that the nests can be identified and harvested with average identification accuracy and harvest success rates of 89.7% and 82%, respectively, with an average harvest time of 4.3 s for different cases. This paper demonstrates the automatic robot system to harvest the edible swiftlet nest through intensive literature review, research, and experiment. Some researchers have pointed out the health benefit of consuming swiftlet nests, but few have looked into developing a robotic system for the harvesting process. Therefore, this study hopes to encourage more renovation in this field of EBN. In this research, the method of harvesting the swiftlet nest mimics how it is actually collected by humans, which uses a scraping tool to separate the nest out of the bird house's wooden frame. First, the swiftlet nest is located by the input from LiDAR's point cloud which is processed to form a 3D representation of the house's cell and segmented to extract the nest portion. As a result, the robot can determine the position of the nests along its planned path, and the scraping end-effector is able to gradually slice the nest from its bottom at a specific angle by our force analysis model. The manipulator is a five DOF arm mounted to a three wheeled mobile platform in which its kinematic is calculated so that the robot's position can be estimated, and the LiDAR is fixed to a one DOF arm to capture the surrounding point cloud at different height levels. The velocity and trajectory of the manipulator movement is controlled based on the calculation at each joint of the arm. The evaluation shows that this robot system is capable of efficiently harvesting the EBN compared to manual action. The total time for the robot to harvest all the six test nests is 113 s, whereas that of human is about 10 min.

Author Contributions: Conceptualization, D.A.V.T. and N.T.T.; methodology, D.A.V.T. and N.T.T.; software, D.A.V.T.; validation, D.A.V.T. and N.T.T.; formal analysis, D.A.V.T.; investigation, D.A.V.T. and N.T.T.; resources, D.A.V.T. and N.T.T.; data curation, D.A.V.T.; writing—original draft preparation, D.A.V.T.; supervision, N.T.T.; project administration, N.T.T.; funding acquisition, N.T.T. All authors have read and agreed to the published version of the manuscript.

Funding: This research is funded by the University of Economics Ho Chi Minh City–UEH University, Vietnam.

Data Availability Statement: No new data were created or analyzed in this study. Data sharing is not applicable to this article.

Conflicts of Interest: The authors declare no conflicts of interest.

Appendix A

$$\cos\theta_2 = \frac{(P_x - l_3\cos\beta)^2 + (P_y - l_3\sin\beta)^2 - l_2^2 - l_3^2}{2l_2l_3}$$

$$\sin\theta_2 = \pm\sqrt{1-\cos^2\theta_2}$$

$$\theta_2 = \text{atan2}(\sin\theta_2, \cos\theta_2)$$

$$\begin{aligned}
&P_x - l_3\cos\beta = l_2\cos\theta_1 + l_3\cos\theta_1\cos\theta_2 - l_3\sin\theta_1\sin\theta_2\\
&= \cos\theta_1(l_2 + l_3\cos\theta_2) + \sin\theta_1(-l_3\sin\theta_2)\\
&P_y - l_3\sin\beta = l_2\sin\theta_1 + l_3\sin\theta_1\cos\theta_2 + l_3\cos\theta_1\sin\theta_2\\
&= \cos\theta_1(l_3\sin\theta_2) + \sin\theta_1(l_2 + l_3\cos\theta_2)\\
&\begin{bmatrix} P_x - l_3\cos\beta \\ P_y - l_3\sin\beta \end{bmatrix} = \begin{bmatrix} l_2 + l_3\cos\theta_2 & -l_3\sin\theta_2 \\ l_3\sin\theta_2 & l_2 + l_3\cos\theta_2 \end{bmatrix}\begin{bmatrix} \cos\theta_1 \\ \sin\theta_1 \end{bmatrix}
\end{aligned}$$

$$\begin{aligned}
\cos\theta_1 &= \frac{\begin{bmatrix} P_x - l_3\cos\beta & -l_3\sin\theta_2 \\ P_y - l_3\sin\beta & l_2 + l_3\cos\theta_2 \end{bmatrix}}{D}\\
&= \frac{(P_x - l_3\cos\beta)(l_2 + l_3\cos\theta_2) + l_3\sin\theta_2(P_y - l_3\sin\beta)}{D}
\end{aligned}$$

$$\begin{aligned}
\sin\theta_1 &= \frac{\begin{bmatrix} l_2 + l_3\cos\theta_2 & P_x - l_3\cos\beta \\ l_3\sin\theta_2 & P_y - l_3\sin\beta \end{bmatrix}}{D}\\
&= \frac{(P_y - l_3\sin\beta)(l_2 + l_3\cos\theta_2) - l_3\sin\theta_2(P_x - l_3\cos\beta)}{D}
\end{aligned}$$

$$\theta_1 = a\tan 2\begin{pmatrix} (P_x - l_3\cos\beta)(l_2 + l_3\cos\theta_2) + l_3\sin\theta_2(P_y - l_3\sin\beta), \\ (P_y - l_3\sin\beta)(l_2 + l_3\cos\theta_2) - l_3\sin\theta_2(P_x - l_3\cos\beta) \end{pmatrix}$$

$$\theta_3 = \beta - \theta_1 - \theta_2$$

$$l_1 = P_z - l_5$$

References

1. Yan, T.H.; Babji, A.S.; Lim, S.J.; Sarbini, S.R. A Systematic Review of Edible Swiftlet's Nest (ESN): Nutritional bioactive compounds, health benefits as functional food, and recent development as bioactive ESN glycopeptide hydrolysate. *Trends Food Sci. Technol.* **2021**, *115*, 117–132. [CrossRef]
2. Dai, Y.; Cao, J.; Wang, Y.; Chen, Y.; Jiang, L. A comprehensive review of edible bird's nest. *Food Res. Int.* **2021**, *140*, 109875. [CrossRef]
3. Mursidah, M.; Lahjie, A.M.; Masjaya, M.; Rayadin, Y.; Ruslim, Y. The ecology, productivity and economic of swiftlet (*Aerodramus fuciphagus*) farming in Kota Bangun, East Kalimantan, Indonesia. *Biodiversitas J. Biol. Divers.* **2020**, *21*, 3117–3126. [CrossRef]
4. El Sheikha, A.F. Why the importance of geo-origin tracing of edible bird nests is arising? *Food Res. Int.* **2021**, *150*, 110806. [CrossRef]
5. Davidson, J.; Bhusal, S.; Mo, C.; Karkee, M.; Zhang, Q. Robotic Manipulation for Specialty Crop Harvesting: A Review of Manipulator and End-Effector Technologies. *Glob. J. Agric. Allied Sci.* **2020**, *2*, 25–41. [CrossRef]
6. Libin, Z.; Yan, W.; Qinghua, Y.; Guanjun, B.; Feng, G.; Yi, X. Kinematicsandtrajectory planning ofacucumber harvestingrobot manipulator. *Int. J. Agric. Biol. Eng.* **2009**, *2*, 1–7. [CrossRef]
7. Vrochidou, E.; Tsakalidou, V.N.; Kalathas, I.; Gkrimpizis, T.; Pachidis, T.; Kaburlasos, V.G. An Overview of End Effectors in Agricultural Robotic Harvesting Systems. *Agriculture* **2022**, *12*, 1240. [CrossRef]
8. Bachche, S.; Oka, K. Performance Testing of Thermal Cutting Systems for Sweet Pepper Harvesting Robot in Greenhouse Horticulture. *J. Syst. Des. Dyn.* **2013**, *7*, 36–51. [CrossRef]
9. Van Henten, E.J.; Hemming, J.; Van Tuijl BA, J.; Kornet, J.G.; Meuleman, J.; Bontsema, J.; Van Os, E.A. An Autonomous Robot for Harvesting Cucumbers in Greenhouses. *Auton. Robot.* **2002**, *13*, 241–258. [CrossRef]
10. Roshanianfard, A.; Noguchi, N. Pumpkin harvesting robotic end-effector. *Comput. Electron. Agric.* **2020**, *174*, 105503. [CrossRef]
11. Mail, M.F.; Maja, J.M.; Marshall, M.; Cutulle, M.; Miller, G.; Barnes, E. Agricultural Harvesting Robot Concept Design and System Components: A Review. *AgriEngineering* **2023**, *5*, 777–800. [CrossRef]
12. Meshram, A.T.; Vanalkar, A.V.; Kalambe, K.B.; Badar, A.M. Pesticide Spraying Robot for Precision Agriculture: A Categorical Literature Review and Future Trends. *J. Field Robot.* **2022**, *39*, 153–171. [CrossRef]
13. Xu, R.; Li, C. A Review of High-Throughput Field Phenotyping Systems: Focusing on Ground Robots. *Plant Phenomics* **2022**, *2022*, 1–20. [CrossRef] [PubMed]
14. Grimstad, L.; Pham, C.D.; Phan, H.T.; From, P.J. On the design of a low-cost, light-weight, and highly versatile agricultural robot. In Proceedings of the 2015 IEEE International Workshop on Advanced Robotics and its Social Impacts (ARSO), Lyon, France, 30 June–2 July 2015; pp. 1–6. [CrossRef]
15. Rong, J.; Wang, P.; Yang, Q.; Huang, F. A Field-Tested Harvesting Robot for Oyster Mushroom in Greenhouse. *Agronomy* **2021**, *11*, 1210. [CrossRef]
16. Mao, W.; Liu, H.; Hao, W.; Yang, F.; Liu, Z. Development of a Combined Orchard Harvesting Robot Navigation System. *Remote Sens.* **2022**, *14*, 675. [CrossRef]
17. Tang, Y.; Chen, M.; Wang, C.; Luo, L.; Li, J.; Lian, G.; Zou, X. Recognition and Localization Methods for Vision-Based Fruit Picking Robots: A Review. *Front. Plant Sci.* **2020**, *11*, 510. [CrossRef] [PubMed]
18. Jiang, S.; Wang, S.; Yi, Z.; Zhang, M.; Lv, X. Autonomous Navigation System of Greenhouse Mobile Robot Based on 3D LiDAR and 2D LiDAR SLAM. *Front. Plant Sci.* **2022**, *13*, 815218. [CrossRef]
19. Wang, L.; Wang, Z.; Liu, M.; Ying, Z.; Xu, N.; Meng, Q. Full Coverage Path Planning Methods of Harvesting Robot with Multi Objective Constraints. *J. Intell. Robot. Syst.* **2022**, *106*, 17. [CrossRef]
20. Goh, D.L.M.; Chua, K.Y.; Chew, F.T.; Liang RC, M.Y.; Seow, T.K.; Ou, K.L.; Yi Bsc, F.C.; Lee, B.W. Immunochemical characterization of edible bird's nest allergens. *J. Allergy Clin. Immunol.* **2001**, *107*, 1082–1088. [CrossRef]
21. Jessel, H.R.; Chen, S.; Osovski, S.; Efroni, S.; Rittel, D.; Bachelet, I. Design principles of biologically fabricated avian nests. *Sci. Rep.* **2019**, *9*, 4792. [CrossRef]
22. Jia, W.; Zhang, Y.; Lian, J.; Zheng, Y.; Zhao, D.; Li, C. Apple harvesting robot under information technology: A review. *Int. J. Adv. Robot. Syst.* **2020**, *17*, 3. [CrossRef]
23. Ramji, S.; Fizl, M.; Koon, L.C.; Rahman, M.A. Roosting and Nest-Building Behaviour of the White-Nest Swiftlet. *Raffles Bull. Zool.* **2013**, *29*, 225–235.

24. Nguyen, V.; Martinelli, A.; Tomatis, N.; Siegwart, R. A Comparison of Line Extraction Algorithms using 2D Laser Rangefinder for Indoor Mobile Robotics. In Proceedings of the Intelligent Robots and Systems (IROS), Edmonton, AL, Canada, 2–6 August 2005. [CrossRef]
25. Yang, C.; Watson, R.M.; Gross, J.N.; Gu, Y. Localization algorithm design and evaluation for an autonomous pollination robot. In Proceedings of the International Meeting of The Satellite Division of the Institute of Navigation, Miami, FL, USA, 16–20 September 2019.

AgriEngineering MDPI

Article

Utilization of Vermicompost Sludge Instead of Peat in Olive Tree Nurseries in the Frame of Circular Economy and Sustainable Development

Vasiliki Kinigopoulou [1,*], Evangelos Hatzigiannakis [1], Stefanos Stefanou [2], Athanasios Guitonas [3] and Efstathios K. Oikonomou [3]

[1] Soil & Water Resources Institute, Hellenic Agricultural Organization "DEMETER", Sindos, 574 00 Thessaloniki, Greece; e.hatzigiannakis@swri.gr
[2] Department of Agriculture, School of Geotechnical Sciences, International Hellenic University, Sindos, 574 00 Thessaloniki, Greece; stefst2@ihu.gr
[3] Faculty of Rural and Surveying Engineering, Aristotle University of Thessaloniki, 541 24 Thessaloniki, Greece; guitonas@topo.auth.gr (A.G.); eoikonom@topo.auth.gr (E.K.O.)
* Correspondence: v.kinigopoulou@swri.gr; Tel.: +30-2310-798790

Abstract: The survival of newly planted seedlings and their successful development after transplantation, including faster plant growth, improved plant quality, larger production, and the absence of dependence on arable land, is one of the primary goals of horticultural nurseries. Although peat is the most frequently used amendment in commercial potting substrates, exploiting it degrades essential ecosystems like peatlands and uses slowly renewable resources. This study evaluated the growth and nutrition of olive-rooted cuttings when peat was partially or completely replaced with vermicompost, searching for more sustainable methods and recovering urban wastewater treatment sludge sequentially. The progress of the plants' growth was compared to that of corresponding plants in which commercial peat had been used as substrate. Leachates from every procedure were also examined, and the results revealed that trace element and heavy metal contents were much lower than those deemed hazardous for aquifers and soil. The outcomes indicated that peat might be effectively replaced with vermicompost sludge, promoting plant growth without further fertilizer. Comparatively to olive cuttings grown in peat-based substrates, those grown in compost-based substrates experienced improved nutrition and development. Further, it was found that irrigation doses were significantly reduced in treatments with a significant amount of vermicompost as the water drained more slowly. A technical-economic analysis was being conducted in the meantime, illustrating the financial benefits for a nursery when peat is replaced with vermicomposted sludge.

Keywords: vermicompost; earthworm compost; peat substitution; olive tree; nursery sustainability; circular economy; feasibility study

Citation: Kinigopoulou, V.; Hatzigiannakis, E.; Stefanou, S.; Guitonas, A.; Oikonomou, E.K. Utilization of Vermicompost Sludge Instead of Peat in Olive Tree Nurseries in the Frame of Circular Economy and Sustainable Development. *AgriEngineering* **2023**, 5, 1630–1643. https://doi.org/10.3390/agriengineering5030101

Academic Editor: Eugenio Cavallo

Received: 23 June 2023
Revised: 12 September 2023
Accepted: 13 September 2023
Published: 19 September 2023

1. Introduction

The most common amendment that is used as a horticultural substrate for growing trees in containers is peat [1]. Since peat is the primary ingredient in growing media for both conventional and organic seedling production, there has been an increase in environmental concern related to its extraction and rising demand, as the use of peat involves the exploitation of slowly renewable resources and the degradation of valuable ecosystems such as peatlands [2]. Therefore, many countries have established several restrictions concerning its use. As a consequence of this practice, many researchers have looked for affordable alternatives to use as environmentally friendly components for substrates. Many organic resources, including residual biomass [3,4], green and municipal wastes, and manure [5–8], have been used as partial or total ingredients in growing media.

Composts may partly replace peat and mineral-based fertilizers while enhancing the properties of the growing medium and lessening the risk of pathogens and parasites, whose presence could be harmful [2,9–12]. Vermicompost is the product generated when earthworms absorb, consume, and decompose a variety of organic wastes, including manure, crop residues, sewage sludge, and industrial wastes [1,2,13,14]. Vermicompost is advantageous to compost because of its high resistance, abundance of nutrients like nitrogen, phosphorus, potassium, calcium, and magnesium, and suitable level of microelements like iron, zinc, copper, and manganese [15]. This conclusion contributes significantly to the fact that vermicomposting increases the humic acid content and emits fewer air pollutants, enhancing microbial biomass potential and enzymatic activities [16]. Enzymes and microorganisms, such as gibberellins, cytokines, auxins, nitrogen-fixing bacteria, mycorrhizae, etc., contained in vermicompost enrich the soil and convert the humic and fulvic acids into humic and fulvic ions, which are readily taken up by the root system and feed the plants.

Earthworm composting is carried out with different species of earthworms, but the species *Eisenia foedita* is best adapted to the climatic conditions of Greece [17]. This earthworm is terrestrial and lives at a depth of less than 10 cm in soil aggregates at temperatures ranging from 10 °C to 30 °C. It dislikes light and produces a cocoon every 10 days, which, after about 15 days, gives 4 to 12 new embryos. Embryos mature in 70 days and double every two to three months at 25 °C, while at 5 °C they double in six months. Earthworms accelerate the ripening of compost, and this is due to two factors: firstly, in their mechanical action, as the excreta of the earthworms stratifies the organic matter, which facilitates, at the end of the ripening of the compost, its separation from the inert non-fermentable materials and therefore the homogenization of the final product. Secondly, in their biological action, earthworms show intense bacterial activity, creating intense trophic competition with pathogenic microorganisms such as salmonella, staphylococci, and enterobacteria (*E. coli*), which disappear during the earthworm composting of 90 days [18,19]. This disappearance is due to simple microbial competition, which reduces the need for substrate nutrition [20].

Meanwhile, the rapid increase in human population has led to a corresponding increase in the amount of wastewater to be treated [21]. Therefore, the amount of sewage sludge produced from wastewater treatment has also increased. During the process of treating wastewater, and especially municipal wastewater, sewage sludge is a by-product containing different nutrients (nitrogen compounds, phosphorus, and potassium) and pollutants such as colloids, pathogenic bacteria, recalcitrant organic compounds, microplastics, etc. [22]. Inadequate handling of sewage sludge could lead to serious environmental harm and several human health problems, such as diarrhea and acute respiratory illnesses. Sewage sludge has been utilized in composting for land filling purposes since its direct field application is not a viable option due to the presence of potentially toxic elements and pathogens, as well as for energy recovery via pyrolysis, incineration, gasification, and methane production [23]. Particularly, although approximately one-third to half of the sewage sludge is recycled for agricultural use, landfilling and incineration are the most common disposal treatments in many countries [24]. According to FAO 2020 [25], sludge treatment and disposal account for 50% of the operational costs of secondary sewage treatment plants in Europe. Raw or treated sewage sludge applied to the soil can meet a considerable portion of the nitrogen and phosphorus needs of many crops while also lowering the cost of sludge disposal in the sewage treatment process. This, combined with the fact that landfilling of all untreated waste, including mixed municipal waste, has been banned [26,27], makes finding sustainable and efficient solutions for sewage treatment a necessity.

The International Olive Council [28] estimates that 43.5 million olive plants are produced worldwide each year. Eighty-four percent of it is produced in the Mediterranean region, with the remaining percentage coming from other nations. Therefore, there could be a significant economic and environmental gain if peat could be replaced by another eco-friendly substrate for nursery olive cultivation. Moreover, to achieve the resource efficiency goals set forth in the Europe 2020 Strategy for smart, sustainable, and inclusive growth [29], it is crucial to move toward a more circular economy.

During this project, sewage sludge stabilization, which is a necessary process to reduce the content of heavy metals and pathogens, took place using earthworms to degrade organic matter and produce a material with very good physical properties, which is odorless and high in essential enzymes, humic acids, microorganisms, and trace elements. The objective of this study was to evaluate the effect of vermicomposted sludge on the growth of young olive trees and to assess the impact of peat replacement with vermicompost, both partially and entirely, on the development and nutrition of olive-rooted cuttings cultivated in a nursery, while searching for integrated, eco-friendly, and financially beneficial solutions and recovering urban wastewater treatment sludge sequentially. Furthermore, a feasibility study took place to evaluate and validate the possible profit of nursery owners, pursuing the objective of sustainable intensification of the circular economy.

2. Materials and Methods

The research was conducted at the olive nursery "Phytogonia" in Kavala, Eastern Macedonia, Greece, during the year 2019. The growth and nutrition of olive-rooted cuttings in a commercial peat-based and non-fertilized substrate were compared with the growth and nutrition in peat-based media substituted with different proportions of vermicompost. The vermicompost was produced from anaerobically composted sewage sludge derived from the wastewater treatment facility "Aeneia" of the Municipality of Thermaikos, Thessaloniki, Greece, using the earthworm *Eisenia foedita*, in the facilities of AKTOR Company, Thessaloniki, Greece, the biggest subsidiary of the ELLAKTOR infrastructure group. The chemical content of peat and vermicompost is presented in Table 1.

Table 1. Chemical characteristics of peat and vermicompost used as potting media.

Parameter	Peat	Vermicompost
OM * (%)	82.2	55.4
pH	5.4	5.6
EC ** (μS cm^{-1})	707	10,130
NH_4-N (mg kg^{-1})	622	288
NO_3-N (mg kg^{-1})	348	21,130
TN *** (g kg^{-1})	9.9	37
C/N	41.5	7.5
P (mg kg^{-1})	441	623
K (mg kg^{-1})	1560	1300
Na (mg kg^{-1})	400	8400
Mg (mg kg^{-1})	728	4430
Ca (mg kg^{-1})	17,166	15,670
B (mg kg^{-1})	2.1	24
Fe (mg kg^{-1})	121	304
Cu (mg kg^{-1})	16.6	42.4
Mn (mg kg^{-1})	121	304
Zn (mg kg^{-1})	17.3	492

* Organic matter, ** electrical conductivity, *** total nitrogen.

2.1. Earthworms Compost Preparation

Recently deposited anaerobically composted sewage sludge was collected from the wastewater treatment facility "Aeneia" of the Municipality of Thermaikos, Thessaloniki. Approximately 4 m^3 of the sewage sludge (water content 85%, organic matter 58%) was placed on the premises of AKTOR Company.

Sewage sludge was accumulated in a high pile and mixed with commercial wheat straw suitable for animal feed in order to adjust the C/N ratio, balance the creamy texture, and absorb, drain, and evaporate gravity and interval water. Since the initial C/N ratio of the sludge was low (≈8.2), a sufficient amount of straw was added to increase the C/N ratio to 22. However, the composting process was sped up as organic matter was already high (58%).

Sewage sludge was mixed twice a week, and the temperature was monitored. Initially, the temperature increased to 65 °C and was maintained at this value for seven days, while subsequently it gradually decreased and stabilized at 45 °C. During these seven days, the pathogens were destroyed and suitable sanitation conditions were created for the earthworm composting, during which the temperature remained lower than 30 °C so that the earthworms did not die.

Then the pile with the compost was transferred, and two rows of dimensions 4 m × 2 m × 0.5 m (length, width, and height) were formed on a concrete platform. The sewage sludge was transferred into containers, and after two weeks, they were vaccinated with earthworms (almost 1 kg m^{-2}). A sprinkler irrigation system was installed on the rows. The rows were irrigated twice a day for 30′ to adjust soil moisture to 65% and stirred twice a week with special shovels. After the earthworms multiplied, their concentration was higher than 2.5 kg m^{-2}. Earthworm composting lasted 90 days, and afterwards, the vermicompost was reduced by about half compared to the sludge. It was then spread on a platform for sun drying in order to minimize the soil moisture from 65% to 25% for pellet production.

2.2. Experimental Treatments

Pilot tests of the produced composted sludge were carried out in an olive-growing nursery. The tests were performed on olive-rooted cuttings of the *Arbequina* olive variety in a mist propagation system. The progress of the plants' growth (height and diameter of the central trunk of the plants) was compared to that of corresponding plants in which commercial peat had been used as substrate. The pilot tests included, in addition to the use of 100% vermicomposted sludge and 100% commercial peat, mixtures of different percentages of the two materials. All of the substrates received a 10% v/v addition of commercial perlite as an inert ingredient, while no fertilizer was added, in order to examine the real effect of the substrate. The proportions (% v/v) of each component regarding each treatment are presented in Table 2. The number of tests was enough to ensure statistical processing of the results obtained. Specifically, *Arbequina* olive-rooted cuttings were transplanted from plug trays under mist propagation conditions into pots (0.6 L) that contained a mixture of commercial peat substrate and different substitutions (0%, 25%, 50%, 75%, and 100%) of vermicompost. Each treatment included 18 replications. Likewise, olive cuttings after the 1st transplanting were transplanted into larger pots (3 L) that also contained a mixture of commercial peat substrate and different substitutions (0%, 25%, 50%, 75%, and 100%) of vermicompost. Each treatment included 12 replications. The potted plants were maintained under nursery conditions, and they were regularly irrigated, depending on shading temperature and solar radiation, to avoid any symptoms of water stress. Chemical analysis of the mixtures was performed, as well as chemical analysis of the leachates from each treatment.

Table 2. Proportions (% v/v) of each component in the growing media *.

Treatment	Vermicompost (%)	Peat (%)
T1	100	-
T2	75	25
T3	50	50
T4	25	75
T5	-	100

* All of the substrates received a 10% v/v addition of commercial perlite as an inert ingredient.

Measurements that took place during this experiment were: chemical and microbial analysis of the partially sun-dried vermicompost and commercial peat used; chemical analysis of all treatments' substrates; chemical analysis of leachates; and a record of plant growth measuring the height of the plants every 24 days as well as the diameter of the plants' trunks. All the chemical analysis took place at the laboratory of the Soil and

Water Resources Institute (SWRI) of the Hellenic Agricultural Organization 'DEMETER' in Sindos, Greece, while microbiological analysis was carried out at the laboratory of the Veterinary Research Institute of the Hellenic Agricultural Organization 'DEMETER' in Thessaloniki, Greece.

2.3. Analytical Methods

Regarding mixtures and leachates, pH was determined using a JENWAY 3520 pH meter, while a CRISON GLP 32 conductimeter was used to assess electrical conductivity (EC) in 1:10 *w*/*v* water-diluted samples. The electrical conductivity (EC) was measured by a CRISON GLP 32 conductimeter in 1:10 *w*/*v* water-diluted samples. The analysis of the organic matter (OM) in the biomixture was performed using the loss-on-ignition method [30]. Using the automatic digital soil calcimeter Fogl, $CaCO_3$ was determined, while a Perkin Elmer Lambda 35 Uv/vIS spectrophotometer was used to colorimetrically measure the extractable amounts of phosphorus (P) and boron (B). After extraction with ammonium acetate, exchangeable potassium (K) and sodium (Na) were measured using a Sherwood M410 flame photometer, while after extraction with ammonium acetate, exchangeable calcium (Ca) and magnesium (Mg) were measured using a Perkin Elmer AAnalyst 200 atomic absorption spectrometer. Following DTPA extraction, the amounts of iron (Fe), copper (Cu), manganese (Mn), and zinc (Zn) were quantified using a Perkin Elmer AAnalyst 200 atomic absorption spectrometer. Based on *Methods of Soil Analysis—Part 3—Chemical Methods*, 1996, Chapter 38 Nitrogen—Inorganic Forms [31], NO_3^- determination was carried out after KCl extraction and after passing through a column of copperized cadmium. NH_4^+ was determined using a Perkin Elmer Lambda 35 UV/VIS spectrophotometer at 667 nm. The total nitrogen (TN) was determined according to ISO 11261:1995, *Soil quality—Determination of total nitrogen—Modified Kjeldahl Method* [32]. Cl was determined by the Mohr titration method [33]. The following equation was used to compute the exchangeable sodium percentage (ESP), which describes the amount of adsorbed Na in soil: ESP = (exchangeable Na/cation exchange capacity) × 100.

2.4. Statistical Analysis

Software entitled SPSS v27 was used to carry out the statistical analysis. For comparison of soil chemical characteristics between the treatments, a one-way ANOVA was conducted. In the present situation, we were looking to test the statistical hypothesis that a mean population set is equal while considering the treatments as a component. The Duncan test ($p < 0.05$) was used to determine whether there were any significant differences between the mean values of the measured parameters.

2.5. Feasibility Study

A technical-economic analysis was carried out in order to investigate the practicality and the possible total profit for an olive nursery owner, taking into account the quantity and quality of the total production, the economic value of peat and vermicompost, whether it is purchased or produced by the nursery owner himself, as well as the partial or total peat substitution by vermicompost, the amount of irrigation water required, and the gain from not using fertilizer throughout the transplanting process.

3. Results and Discussion

3.1. Experimental Outcomes

The partial and total substitution of peat by vermicompost increased the height and the diameter of the olive-rooted cuttings at both the 1st and 2nd transplantings (Figures 1 and 2). These outcomes were quite encouraging, contrary to Regni et al. [34], who evaluated the impact of peat substitution by new-type olive mill pomace and its compost on the vegetative activity of potted olive trees and came to the conclusion that up to 50% of peat could be replaced with compost without substantially decreasing the development of plants in the end. It was also observed that the height of the plants in the T2 (VC-75%) treatment was almost twice the height of the T5 (P-100%) treatment after the 1st transplanting (Figure 3). Similarly, after the 2nd transplanting, all the treatments with vermicompost were much more effective than T5 substrate. These results were rather positive compared to the observations by Tosca et al. [35], who replaced peat with biochar and green compost for olive tree cultivation, and no significant effect on plant height was detected, whereas branching was slightly diminished in some treatments. As concluded by the experimental outcomes, cuttings of the compost-based substrates grew better than olive cuttings grown in the peat-based substrate, especially when the mix composition was 75% compost–25% peat (T2) (Figures 3 and 4). At this point, it is worth noting that the materials of the treatments VC-100% presented inhomogeneity in their composition, shrinkage into the pots, as well as difficulty in draining.

The results of the chemical analysis of the different growing media after the 1st and 2nd transplanting are presented in Tables 3 and 4, while in Tables 5 and 6, the concentration range of chemical parameters of the treatments' leachates after the 1st and 2nd transplanting and during the irrigation scheme are demonstrated.

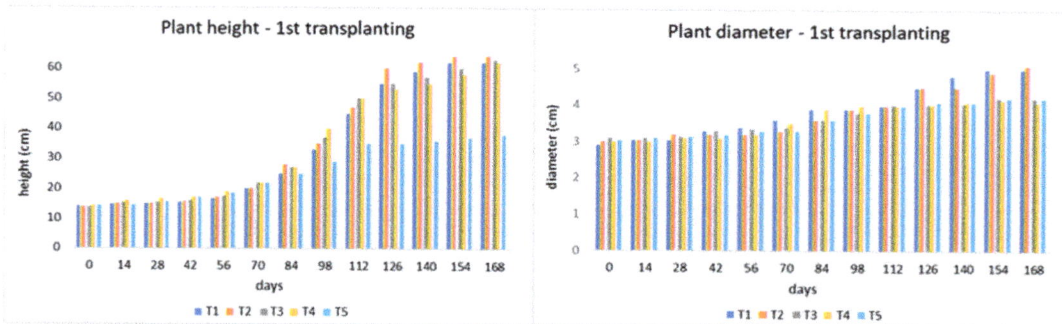

Figure 1. Plant height and diameter of each treatment after 1st transplanting. Blue column: T1 (100% vermicompost), orange column: T2 (75% vermicompost–25% peat), grey column: T3 (50% vermicompost–50% peat), yellow column: T4 (25% vermicompost–75% peat), light blue column: T5 (100% peat).

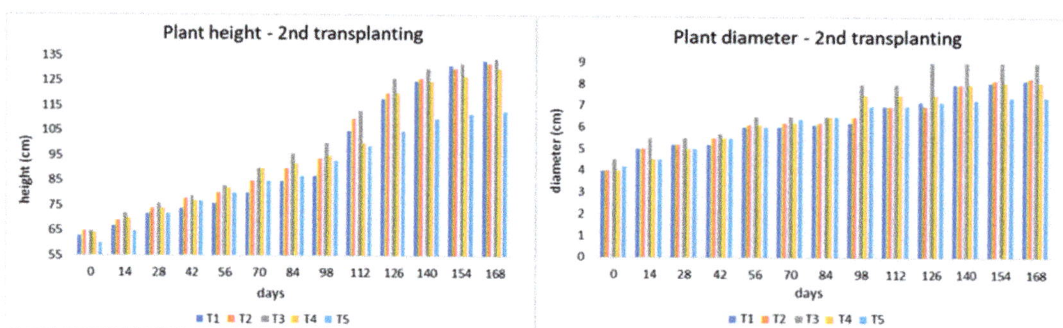

Figure 2. Plant height and diameter of each treatment after 2nd transplanting. Blue column: T1 (100% vermicompost), orange column: T2 (75% vermicompost–25% peat), grey column: T3 (50% vermicompost–50% peat), yellow column: T4 (25% vermicompost–75% peat), light blue column: T5 (100% peat).

Figure 3. Comparison of plant growth between T5 and T2 treatments after the 1st transplanting.

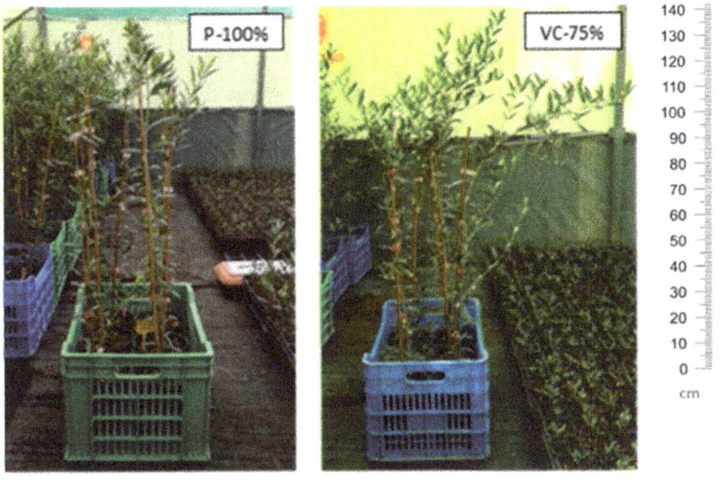

Figure 4. Comparison of plant growth between T5 and T2 treatments after the 2nd transplanting.

Table 3. Chemical parameters of substrates in the different treatments after the 1st transplanting.

Treatment	T1	T2	T3	T4	T5
pH	6.5 a	6.1 b	5.8 c	5.5 d	4.9 e
EC * (μS cm^{-1})	8021 a	7575 b	6521 c	3142 d	2259 e
OM ** (%)	54.2 e	60.5 d	66.9 c	75.9 b	87.6 a
C/N	7.8 e	10.9 d	13.4 c	18.6 b	37.3 a
K (mg kg^{-1})	1204 e	1456 d	1795 c	2208 b	2300 a
Na (mg kg^{-1})	8171 a	7583 b	6151 c	3300 d	206 e
P (mg kg^{-1})	462 d	467 d	501 c	490 b	591 a
Ca (mg kg^{-1})	6105 e	6905 d	7475 c	8515 b	9388 a
Mg (mg kg^{-1})	1980 b	2030 a	1520 c	1365 d	972 e

Table 3. *Cont.*

Treatment	T1	T2	T3	T4	T5
N-NO_3 (mg kg^{-1})	1451 a	1299 b	1314 b	1224 c	1038 d
N-NH_4 (mg kg^{-1})	108.4 cd	98.9 d	117.7 c	190.6 b	256.9 a
TN (g kg^{-1})	34.7 a	27.8 b	25.3 c	20.5 d	11.9 e
B (mg kg^{-1})	23.2 a	22.7 a	22.2 a	15.2 b	1.2 c
Cu (mg kg^{-1})	43.7 a	36.2 b	34.6 b	21.1 c	10.5 d
Zn (mg kg^{-1})	466 a	390 b	325 c	167 d	18.1 e
Fe (mg kg^{-1})	315 a	255 b	241 b	246 b	183 c
Mn (mg kg^{-1})	73.6 a	72.8 a	75.2 a	75.0 a	48.6 b

Mean values in the same row that are succeeded by different letters vary considerably ($p < 0.05$). * Electrical conductivity, ** organic Matter.

Table 4. Chemical parameters of substrates in the different treatments after the 2nd transplanting.

Treatment	T1	T2	T3	T4	T5
pH	6.5 a	6.1 b	5.7 c	5.7 c	5.4 d
EC * (μS cm^{-1})	7037 a	5878 b	5338 c	4042 d	987 e
OM ** (%)	47.2 d	47.3 d	53.3 c	59.0 b	67.2 a
C/N	7.29 e	9.49 d	12.0 c	24.3 b	45.2 a
K (mg kg^{-1})	1000 b	950 c	1050 a	1000 b	1000 b
Na (mg kg^{-1})	6750 a	5400 b	4396 c	2650 d	180 e
P (mg kg^{-1})	445 a	408 b	376 c	318 d	165 e
Ca (mg kg^{-1})	6048 c	5310 d	6015 c	6200 b	6960 a
Mg (mg kg^{-1})	1890 a	1550 b	1495 c	1060 d	490 e
N-NO_3 (mg kg^{-1})	1128 a	874 b	796 c	639 d	259 e
N-NH_4 (mg kg^{-1})	97.1 c	78.0 d	74.3 d	109.2 b	173.0 a
TN (g kg^{-1})	32.8 a	25.4 b	22.9 b	12.3 c	7.5 d
B (mg kg^{-1})	21.4 a	19.5 b	17.2 c	12.4 d	2.2 e
Cu (mg kg^{-1})	38.6 a	33.9 b	32.2 b	18.6 c	11.4 d
Zn (mg kg^{-1})	395 a	306 b	271 b	161 c	16.3 d
Fe (mg kg^{-1})	262 ab	272 a	247 b	228 c	180 d
Mn (mg kg^{-1})	65.8 b	71.8 ab	75.2 a	64.5 b	33.4 c

Mean values in the same row that are succeeded by different letters vary considerably ($p < 0.05$). * Electrical conductivity, ** organic matter.

Table 5. Concentration range of chemical parameters of the treatments' leachates after the 1st transplanting during the irrigation scheme.

Treatment	T1	T2	T3	T4	T5
pH	7.3–8.8	7.1–9.1	7.0–8.8	6.6–8.3	6.6–7.7
EC * (mS cm^{-1})	4.5–1.1	6.5–1.2	4.6–1.3	4.5–1.4	1.9–1.1
NO_3 (mg L^{-1})	1446–33.3	1551–48.3	1185–23.6	1401–20.5	301–16.2
NH_4 (mg L^{-1})	0.2–3.3	0.3–2.1	0.2–2.2	0.6–3.1	0.6–1.5
K (mg kg^{-1})	41–4.7	80–2.8	80–2.3	115–2.5	125–5.6
Na (mg kg^{-1})	415–70	607–105	364–140	301–132	139–63
P (mg kg^{-1})	5.9–0.8	9.4–1.2	11.0–2.3	23.5–3.5	30.8–0.1
Ca (mg kg^{-1})	354–127	529–129	403–136	504–140	210–99
Mg (mg kg^{-1})	142–32.1	200–28.6	143–32.5	134–30.8	39.0–25.0
Cl (mg L^{-1})	602–49.6	889–89.5	493–130	362–115	276–71.0
$CaCO_3$ (mg L^{-1})	1468–448	2144–441	1596–473	1807–477	681–350
SAR **	4.69–1.43	5.68–2.14	3.95–2.27	3.27–2.49	2.67–1.07
SO_4 (mg L^{-1})	352–12.3	659–34.3	779–14.5	769–11.4	1103–80.5
B (mg L^{-1})	0.25–0.16	0.09–0.18	0.07–0.16	0.06–0.31	0.08–0.25
Cu (μg L^{-1})	70–26	78–28	76–17	64–28	69–16
Zn (mg L^{-1})	0.20–0.10	0.20–0.10	0.20–0.10	0.20–0.10	0.20–0.10
Fe (μg L^{-1})	21–70	17–78	20–63	40–77	40–388
Mn (μg L^{-1})	218–32	554–50	497–15	576–49	141–40

* Electrical conductivity, ** sodium adsorption rate.

Table 6. Concentration range of chemical parameters of the treatments' leachates after the 2nd transplanting during the irrigation scheme.

Treatment	T1	T2	T3	T4	T5
pH	7.5–7.8	6.8–7.5	6.5–7.5	6.6–7.8	6.1–7.6
EC * (mS cm^{-1})	10.0–1.6	13.6–2.7	7.8–2.4	4.9–1.5	2.5–1.0
NO_3^- (mg L^{-1})	961–153	2380–584	2285–362	1533–114	859–16
NH_4^+ (mg L^{-1})	70–0.2	42–0.3	5.8–0.3	3.4–0.5	4.5–0.3
K (mg kg^{-1})	108–5.3	160–3.0	95–2.8	85–2.0	122–2.5
Na (mg kg^{-1})	1113–119	1467–190	670–195	341–147	200–56
P (mg kg^{-1})	6.6–2.2	11.6–4.6	22.4–6.5	37.3–4.1	33.4–0.3
Ca (mg kg^{-1})	500–174	589–317	676–298	506–178	312–147
Mg (mg kg^{-1})	447–52.2	496–113	237–65.5	120–51.4	49.0–26.3
Cl (mg L^{-1})	1639–89.1	2442–205	995–263	518–155	156–53.2
$CaCO_3$ (mg L^{-1})	3090–650	3512–1262	2665–1140	1757–671	953–475
SAR **	8.67–1.53	10.7–2.32	5.63–2.50	3.71–2.44	2.82–0.79
SO_4^{2-} (mg L^{-1})	2162–207	736–188	698–38	498–30	407–170
B (mg L^{-1})	0.48–0.43	0.66–0.35	0.64–0.23	0.18–0.22	0.03–0.25
Cu (μg L^{-1})	65–18	89–20	75–17	68–14	82–15
Zn (mg L^{-1})	0.20–0.10	0.58–0.10	0.63–0.10	0.85–0.10	0.32–0.10
Fe (μg L^{-1})	18–274	30–206	50–100	30–150	27–1600
Mn (μg L^{-1})	337–33	773–56	1081–62	949–59	429–65

* Electrical conductivity, ** sodium adsorption rate.

Based on the statistical analysis of the outcomes presented in Tables 3 and 4, it is observed that after the 1st transplanting, the highest values of pH and EC as well as the highest concentrations of TN, N-NO_3, Na, B, Cu, Zn, and Fe were recorded in T1 treatment, while the highest values of C/N and the highest concentrations of OM, N-NH_4, P, K, and Ca were recorded in T5 treatment. Except for the case of P, these results are in accordance with the chemical characteristics of peat and vermicompost presented in Table 1. Concerning the measurements after the 2nd transplanting, similar trends to those of the 1st transplanting were recorded except for the case of P, whose highest concentration was measured in T1 treatment instead of T5, and the case of K, which showed the same concentrations in both T1 and T5 treatments.

It has been ascertained that compost made from green leftovers works well as part of growing media [36]. Particularly, it was discovered that adding small amounts of green compost made from municipal biosolids to a peat-based medium might significantly improve the growth of olive seedlings [37]. Meanwhile, in order to assess the potential of green nursery compost to replace different percentages of peat in potting mixes used for growing olive pot plants, the nitrogen mineralization ability and the identification of olive tree growth parameters were utilized. The results showed that green compost may serve as an appropriate portion of mixed-peat substrates for olive trees, delivering a 15% or 30% peat replacement [38]. Our outcomes proved that peat might be effectively replaced by vermicomposted sludge, promoting the development of plants without further fertilizer. As the substrate becomes favorably less acidic compared to olive cuttings developing in a peat-based medium, cuttings of the compost-based mediums received adequate nutrients, with OM content, C/N ratio, P, K, and Ca concentrations in peat-vermicompost mixtures being quite sufficient, while Na, Mg, N-NO_3, TN, B, Cu, Zn, Fe, and Mn content reaching higher values than those in peat substrate. Specifically, K content after the 1st and 2nd transplanting of T2, T3, and T4 treatments ranged between 1000 and 2000 mg kg^{-1}, and P content after the 1st and 2nd transplanting of T2, T3, and T4 treatments ranged between 300 and 500 mg kg^{-1}. Moreover, the concentrations of NO_3-N in treatments T2, T3, and T4 are lower than in vermicompost, and in fact, in all treatments of the 2nd transplanting, they are even lower than in the 1st transplanting, which is attributed to the absorption of nitrate nitrogen by the plants as well as its leaching and the fact that the rate at which nitrification occurs typically decreases as substrate pH reduces [39]. Actually, the procedure of nitrification begins when the pH of the substrates is more than 6.0 [40]. Meanwhile, the high Na content of vermicompost is reflected in all the treatments, as is the boron, copper, zinc, and iron content.

Concerning the leachates' chemical characteristics, the outcomes indicated that their nutrient content was quite high, indicating adequate coverage of nutrient needs, while the amounts of heavy metals were significantly lower than those deemed dangerous for soil and groundwater (Table 7). Initially, soil, peat, and vermicompost contained some heavy metals in small concentrations, which were further reduced during mixing, apparently due to their complexation with organic matter. Also worth noting is the fact that low boron and zinc concentrations were observed in the leachates, apparently due to both selective assimilation by plants and the low mobility of these elements, indicating that there is a reserve of these trace elements as they are retained in the substrate, which is an advantage for olive cultivation [41,42].

Table 7. Measured values of metals in the leachates of both transplantings during the irrigation scheme compared to acceptable value limits [z] [43].

Metals	Measured Values		Acceptable Value Limits
	Min	Max	
As ($\mu g\ L^{-1}$)	<1.0	<1.0	2000
Cd ($\mu g\ L^{-1}$)	<0.3	16.0	50.0
Cr ($\mu g\ L^{-1}$)	<2.0	<2.0	1000
Cu ($\mu g\ L^{-1}$)	<10.0	89.0	5000
Fe ($\mu g\ L^{-1}$)	17.0	1606	20,000
Mn ($\mu g\ L^{-1}$)	<6.52	1081	10,000
Ni ($\mu g\ L^{-1}$)	<5.0	1268	2000
Zn ($\mu g\ L^{-1}$)	<250	848	10,000
Hg ($\mu g\ L^{-1}$)	<1.0	<1.0	1.0

[z] Values are the mean of three replicates.

Regarding the results of the microbiological analyses, it appears that the produced vermicomposted sludge is a sanitized product, with the absence of salmonella/25 g and *E. coli* content of 10^3/g, which is below the limit value of 2×10^6 CFU/g DS of the Class B pathogen requirement set forth by the U.S. Environmental Protection Agency (EPA) [44], while the anaerobic sludge showed the presence of salmonella/25 g and *E. coli* values of 10^6/g, which are very high. This fact indicated that pathogens had been diminished to values that preserved both nature and public health, and strategies for using sludge involved limitations on agriculture harvesting, livestock grazing, and public engagement [45].

3.2. Feasibility Study Outcomes

Regarding the financial aspect, the annual cost reduction was calculated in the specific olive nursery where the experiment was carried out, which produces 260,000 plants of the 1st transplanting in pots with a volume of 0.6 L and 40,000 large plants of the 2nd transplanting in pots with a volume of 3 L.

The treatment that gave the best results was T2 (75% vermicompost–25% peat). Therefore, in the 260,000 pots with a volume of 0.6 L, 117,000 L (260,000 $\times$ 0.6 $\times$ 0.75) of peat were replaced by vermicomposted sludge, while in the 40,000 pots with a volume of 3 L, 90,000 L (40,000 $\times$ 3 $\times$ 0.75) of peat were replaced by earthworm composted sludge. That is, in this nursery, 117,000 + 90,000 = 207,000 L of peat were replaced by earthworm-composted sludge. Nowadays, the cost of peat amounts to 0.1 EUR/L.

Vermicomposting of the anaerobically composted sewage sludge costs 30 EUR/ton when it takes place at the Wastewater Treatment Facility, while the transport costs amount to 10 EUR/ton. The price of the vermicomposted sludge ranges from 110 to 120 EUR/ton, as long as it contains all the necessary nutrients. Considering that the vermicomposted sludge has a bulk density value of about 0.9 ton/m^3, so the ton is about 1100 L, the purchase cost of the vermicomposted sludge is 0.1 EUR/L, similar to peat. On the other hand, when vermicomposted sludge is produced at the nursery, the cost amounts to 0.035 EUR/L. Thus, in this case, there is a significant profit of $0.065 \times 207{,}000$ = EUR 13,455 per year for the nursery, while in the case of purchasing the vermicompost, there is no profit.

Irrigation of the plants is considered necessary throughout the year, but it varies every month as it depends on the prevailing weather conditions. According to the data from 2019 presented in Table 8, we found that the cost of irrigation for the months with the highest temperatures was EUR 97 per month. Therefore, it was estimated that for 300,000 plants, the irrigation cost in 2019 amounted to EUR 711.65. From the results of the T2 treatment, we conclude that by replacing the peat with VC-75%, we achieve a reduction in irrigation needs of approximately 30%, so the total cost of irrigation for a year is reduced to EUR 498.16. This reduction is attributed to the soil structure, to which a higher percentage of vermicompost is added, leading to greater water retention.

Table 8. Irrigation cost per month for year 2019.

Month	Irrigation Cost (EUR)	Irrigations/Month	Cost (EUR)/Month
January	6.67	2	12.94
February	6.67	3	19.41
March	6.67	8	51.76
April	6.67	10	64.70
May	6.67	15	97.05
June	6.67	15	97.05
July	6.67	15	97.05
August	6.67	15	97.05
September	6.67	12	77.64
October	6.67	8	51.76
November	6.67	4	25.88
December	6.67	3	19.41
Total		110	711.65

Regarding fertilization, the usual fertilizer used in the nursery costs 109 EUR/25 kg. Standard nursery practice is to use 2 kg of fertilizer per m^3 of substrate. So, it was calculated that for about 300,000 plants ($260{,}000 \times 0.6$ L + $40{,}000 \times 3$ L = 276 m^3), 552 kg of fertilizer is needed. Then, the total cost for fertilization amounts to EUR 2406.72.

Therefore, vermicomposted sludge is a substrate that can act as a substitute for peat and fertilizer; it is easy to prepare; it is environmentally friendly; and at the same time, it offers a reduction in the cost of olive tree production. Specifically, the reduction in costs is presented in Table 9.

Table 9. Cost reduction for year 2019.

	Cost Reduction by	When Vermicompost Is Produced at the Nursery	When Vermicompost Is Purchased
✓	Use of vermicompost	EUR 13,455	EUR 0
✓	Irrigation reduction	EUR 498.16	EUR 498.16
✓	No fertilization	EUR 2406.72	EUR 2406.72
	Total	EUR 16,359.88	EUR 2905.88

To summarize, regarding the technical-economic analysis that took place, it was concluded that the annual cost reduction for an olive tree nursery that produces about 300,000 olive trees, in case the peat is substituted by a mixture of 75% vermicompost–25% peat as a growing medium, will be around EUR 2500 if the vermicompost is purchased commercially, while the annual reduction will be around EUR 16,000 if the compost is produced at the nursery. This reduction lies both in the replacement of peat by vermicompost and the non-use of fertilizer throughout the production process and in the reduced irrigation needs, while it was noticed that the irrigation dosages were much lower in the treatments that had a high proportion of vermicompost as the water drained more slowly.

4. Conclusions

Due to ecological restrictions linked to peat use, peat replacement is becoming more and more necessary in the production of pots and transplants. Vermicomposting and using the final product to partially replace peat in nursery activities allow for a reduction in the financial and environmental costs associated with the production of potted plants. This work evaluated the utilization of vermicomposted sludge as an alternate growing medium instead of peat for the transplanting of olive-rooted cuttings and young trees in olive tree nurseries. The results of partial and total peat substitution experiments indicated that cuttings of the compost-based media obtained greater development and nutrient levels in comparison to olive-rooted cuttings grown in peat-based media, especially when the mix composition was 75% compost–25% peat, while the irrigation doses were found to be much lower in treatments that contained a high percentage of vermicomposted sludge because the water drained more slowly. So, minimized irrigation needs, in combination with the fact that plant growth was enhanced while no fertilization took place, pointed out the potential economic benefit arising from peat substitution by earthworm compost sludge for the owner of the nursery. On the other hand, the materials of the vermicompost 100% presented inhomogeneity in their composition, shrinkage into the pots, as well as difficulty in draining. Also, leachates from each treatment were analyzed, and the findings demonstrated that trace element and heavy metal content were much below those regarded as dangerous for aquifers. In conclusion, the observed boost in plant quality makes the use of vermicomposted sludge an advantageous substitute for the use of peat in olive tree nurseries, but also because of the associated economic and ecological advantages, as reusing and employing urban wastewater treatment plants' sludge is a cost-effective and sustainable approach. Further research is needed to optimize the vermicomposting process by including other parameters to be investigated, such as root length, number of leaves, leaf area, etc. Also, further investigation is needed to reduce pathogens and assess the vermicomposting effect on older olive trees.

Author Contributions: Conceptualization, E.H. and A.G.; methodology, E.H.; software, V.K. and S.S.; validation, E.H., A.G. and E.K.O.; formal analysis, V.K.; investigation, V.K. and E.H.; resources, A.G.; data curation, S.S.; writing—original draft preparation, V.K.; writing—review and editing, V.K. and E.H.; visualization, S.S.; supervision, A.G.; project administration, A.G.; funding acquisition, A.G. and E.K.O. All authors have read and agreed to the published version of the manuscript.

Funding: This research was supported by the Soil and Water Resources Institute of the Hellenic Agricultural Organization "DEMETER", and the School of Rural and Surveying Engineering, Aristotle University of Thessaloniki, Greece.

Data Availability Statement: All data are provided within the paper.

Acknowledgments: The authors thank D. Paraskevas for providing the nursery and assisting in the experimental progress.

Conflicts of Interest: The authors declare no conflict of interest.

References

1. Sax, M.S.; Scharenbroch, B.C. Assessing Alternative Organic Amendments as Horticultural Substrates for Growing Trees in Containers. *J. Environ. Hortic.* **2017**, *35*, 66–78. [CrossRef]
2. Bustamante, M.A.; Gomis, M.P.; Pérez-Murcia, M.D.; Gangi, D.; Ceglie, F.G.; Paredes, C.; Pérez-Espinosa, A.; Bernal, M.P.; Moral, R. Use of livestock waste composts as nursery growing media: Effect of a washing pre-treatment. *Sci. Hortic.* **2021**, *281*, 109954. [CrossRef]

3. Cocozza, C.; Parente, A.; Zaccone, C.; Mininni, C.; Santamaria, P.; Miano, T. Chemical, physical and spectroscopic characterization of *Posidonia oceanica* (L.) Del. residues and their possible recycle. *Biomass Bioenergy* **2011**, *35*, 799–807. [CrossRef]
4. Andreu-Rodriguez, J.; Medina, E.; Teresa Ferrandez-Garcia, M.; Ferrandez-Villena, M.; Eugenia Ferrandez-Garcia, C.; Paredes, C.; Bustamante, M.A.; Moreno-Caselles, J.; Moreno, J. Agricultural and Industrial Valorization of *Arundo donax* L. *Commun. Soil Sci. Plant Anal.* **2013**, *44*, 598–609. [CrossRef]
5. Ceglie, F.G.; Bustamante, M.A.; Amara, M.B.; Tittarelli, F. The Challenge of Peat Substitution in Organic Seedling Production: Optimization of Growing Media Formulation through Mixture Design and Response Surface Analysis. *PLoS ONE* **2015**, *10*, e0128600. [CrossRef]
6. Sáez, J.A.; Clemente, R.; Bustamante, M.Á.; Yañez, D.; Bernal, M.P. Evaluation of the slurry management strategy and the integration of the composting technology in a pig farm—Agronomical and environmental implications. *J. Environ. Manag.* **2017**, *192*, 57–67. [CrossRef]
7. Morales, A.B.; Ros, M.; Ayuso, L.M.; De Los, M.; Bustamante, A.; Moral, R.; Pascual, J.A. Agroindustrial composts to reduce the use of peat and fungicides in the cultivation of muskmelon seedlings. *J. Sci. Food Agric.* **2016**, *97*, 875–881. [CrossRef]
8. Jara-Samaniego, J.; Pérez-Murcia, M.D.; Bustamante, M.A.; Pérez-Espinosa, A.; Paredes, C.; López, M.; López-Lluch, D.B.; Gavilanes-Terán, I.; Moral, R. Composting as sustainable strategy for municipal solid waste management in the Chimborazo Region, Ecuador: Suitability of the obtained composts for seedling production. *J. Clean. Prod.* **2017**, *141*, 1349–1358. [CrossRef]
9. Gocmez, S.; Belliturk, K.; Gorres, J.H.; Turan, H.S.; Ustundag, O. The Effects of the Use of Vermicompost in Olive Tree Farming on Microbiological and Biochemical Characteristics of the Production Material. *Erwerbs-Obstbau* **2019**, *61*, 337–344. [CrossRef]
10. Ferraz Ramos, R.; Almeida Santana, N.; de Andrade, N.; Scheffer Romagna, I.; Tirloni, B.; de Oliveira Silveira, A.; Domínguez, J.; Josemar Seminoti Jacques, R. Vermicomposting of cow manure: Effect of time on earthworm biomass and chemical, physical, and biological properties of vermicompost. *Bioresour. Technol.* **2022**, *345*, 126572. [CrossRef]
11. Khomami, A.M.; Haddad, A.; Alipoor, R.; Hojati, S.I. Cow manure and sawdust vermicompost effect on nutrition and growth of ornamental foliage plants. *Cent. Asian J. Environ. Sci. Technol. Innov.* **2021**, *2*, 68–78.
12. Meza-Figueroa, C.A.; Gualdron-Linares, D.G.; Ramírez-Guerrero, H.O. Using compost and vermicompost for enrichment of alternative local substrate medium in bell pepper transplant production. *Acta Hortic.* **2019**, *1249*, 135–140. [CrossRef]
13. Alidadi, H.; Saffari, A.R.; Ketabi, D.; Peiravi, R.; Hosseinzadeh, A. Comparison of Vermicompost and Cow Manure Efficiency on the Growth and Yield of Tomato Plant. *HealthnScope* **2014**, *3*, e14661. [CrossRef]
14. Bajsa, O.; Nair, J.; Mathew, K.; Ho, G.E. Vermiculture as a tool for Domestic Wastewater management. *Water Sci. Technol.* **2003**, *48*, 125–132. [CrossRef] [PubMed]
15. Hosseinzadeh, S.R.; Amiri, H.; Ismaili, A. Effect of vermicompost fertilizer on photosynthetic characteristics of chickpea (*Cicer arietinum* L.) under drought stress. *Photosynthetica* **2016**, *54*, 87–92. [CrossRef]
16. Lakhdar, A.; Rabhi, M.; Ghnaya, T.; Montemurro, F.; Jedidi, N.; Abdelly, C. Effectiveness of compost use in salt-affected soil. *J. Hazard. Mater.* **2009**, *171*, 29–37. [CrossRef]
17. Papaconstantinou, S.; Guitonas, A. *Technical Annals*; Technical Chamber of Greece: Thessaloniki, Greece, 1988; pp. 81–92.
18. Monroy, F.; Aira, M.; Dominguez, J. Changes in density of nematodes, protozoa and total coliforms after transit through the gut of four epigeic earthworms (*Oligochaeta*). *Appl. Soil Ecol.* **2008**, *39*, 127–132. [CrossRef]
19. Ganesh Kumar, A.; Sekaran, G. Enteric Pathogen modification by Anaecic Earthworm, Lampito Mauritii. *J. Appl. Sci. Environ. Manag.* **2005**, *9*, 15–17.
20. Loquet, M.; Vinceslas, M.; Rouelle, J. Cellulasic activity in the gut of *Eisenia foetida*. *Appl. Biochem. Biotechnol.* **1984**, *9*, 377. [CrossRef]
21. Aydın, S.; Ulvi, A.; Bedük, F.; Aydın, M.E. Pharmaceutical residues in digested sewage sludge: Occurrence, seasonal variation and risk assessment for soil. *Sci. Total Environ.* **2022**, *817*, 152864. [CrossRef]
22. Yu, L.; Zhang, Y.; Zhang, Z.; Mao, H.; Han, H.; Yang, J. Recycling reuse of municipal sewage sludge in sustainable structural materials: Preparation, properties, crystallization and microstructure analyses. *Constr. Build. Mater.* **2023**, *398*, 132507. [CrossRef]
23. Al Samsami, O.; Etri, T.; Baawain, M. Characterization of sewage sludge in Muscat and estimation of their energy recovery. *Sustain. Energy Technol. Assess.* **2023**, *58*, 103351. [CrossRef]
24. Aleisa, E.; Alsulaili, A.; Almuzaini, Y. Recirculating treated sewage sludge for agricultural use: Life cycle assessment for a circular economy. *Waste Manag.* **2021**, *135*, 79–89. [CrossRef] [PubMed]
25. Dreyer, H. *FAO: AGP—Fertilizer Specifications*; The United Nations, Food and Agriculture Organization of the United Nations: Rome, Italy, 2020.
26. EEA. *Diverting Waste from Landfill: Effectiveness of Waste-Management Policies in the European Union*; European Environment Agency: Copenhagen, Denmark, 2009.
27. Przydatek, G.; Wota, A.K. Analysis of the comprehensive management of sewage sludge in Poland. *J. Mater. Cycles Waste Manag.* **2020**, *22*, 80–88. [CrossRef]
28. International Olive Council. *Olive Nursery Production and Plant Production Techniques*; International Olive Council: Madrid, Spain, 2015.
29. European Commission. Europe 2020: A Strategy for Smart, Sustainable and Inclusive Growth: Communication from the Commission. Publications Office of the European Union. Work. Pap. {COM 2020}. 2010. Available online: http://eurlex.europa.eu/LexUriServ/LexUriServ.do?uri=COM:2010:2020:FIN:EN:PDF (accessed on 15 June 2023).

30. Ben-Dor, E.; Banin, A. Determination of organic matter content in arid-zone soils using a simple "loss-on-ignition" method. *Commun. Soil Sci. Plant Anal.* **1989**, *20*, 1675–1695. [CrossRef]
31. Keeney, D.R.; Nelson, D.W. *Methods of Soil Analysis*; Page, A.L., Ed.; Agronomy Monographs; American Society of Agronomy, Soil Science Society of America: Madison, WI, USA, 1983; ISBN 9780891189770.
32. *ISO 11261:1995*; Soil Quality—Determination of Total Nitrogen—Modified Kjeldahl Method. ISO: Geneva, Switzerland, 1995. Available online: https://www.iso.org/standard/19239.html (accessed on 18 February 2021).
33. Belcher, R.; Macdonald, A.M.G.; Parry, E. On mohr's method for the determination of chlorides. *Anal. Chim. Acta* **1957**, *16*, 524–529. [CrossRef]
34. Regni, L.; Pezzolla, D.; Gigliotti, G.; Proietti, P. The sustainable reuse of compost from a new type of olive mill pomace in replacing peat for potted olive tree. *Agron. Res.* **2020**, *18*, 1440–1454.
35. Tosca, A.; Valagussa, M.; Martinetti, L.; Frangi, P. Biochar and green compost as peat alternatives in the cultivation of photinia and olive tree. *Acta Hortic.* **2021**, *1305*, 257–262. [CrossRef]
36. Benito, M.; Masaguer, A.; Moliner, A.; Hontoria, C.; Almorox, J. Dynamics of pruning waste and spent horse litter co-composting as determined by chemical parameters. *Bioresour. Technol.* **2009**, *100*, 497–500. [CrossRef]
37. Mugnai, S.; Azzarello, E.; Mancuso, S.; Pandolfi, C.; Masi, E. Investigating the possibility of peat substitution in olive nurseries by green compost. *Adv. Hortic. Sci.* **2007**, *21*, 96–100.
38. Aleandri, M.P.; Chilosi, G.; Muganu, M.; Vettraino, A.; Marinari, S.; Paolocci, M.; Luccioli, E.; Vannini, A. On farm production of compost from nursery green residues and its use to reduce peat for the production of olive pot plants. *Sci. Hortic.* **2015**, *193*, 301–307. [CrossRef]
39. Benke, M.B.; Goh, T.B.; Karamanos, R.; Lupwayi, N.Z.; Hao, X. Retention and nitrification of injected anhydrous NH_3 as affected by soil pH. *Can. J. Soil Sci.* **2012**, *92*, 589–598. [CrossRef]
40. Focht, D.D. Biochemical Ecology of Nitrification and Denitrification. In *Advances in Microbial Ecology*; Alexander, M., Ed.; Plenum: New York, NY, USA, 1977; Volume 1, pp. 135–214.
41. Jan, M.; Sumrah, M.A.; Akhtar, S.; Haq, I.U.; Ramzan, M.; Yasmin, I. Nutritional requirement of olive (*Olea europaea* L.) in Pothwar Region of Pakistan: A review. *Sarhad J. Agric.* **2021**, *37*, 1334–1341. [CrossRef]
42. Akdemir, B.; Saglam, C.; Belliturk, K.; Makaraci, A.; Urusan, A.; Atar, E. Effect of spatial variability on fertiliser requirement of olive orchard cultivated for oil production. *J. Environ. Prot. Ecol.* **2018**, *19*, 319–329.
43. Panoras, A.; Ilias, A. *Irrigation with Treated Urban Waste*, 1st ed.; GIAHOUDI-GIAPOULI: Thessaloniki, Greece, 1999; ISBN 960-91087-0-9.
44. U. S. Environmental Protection Agency. Standards for use or disposal of sewage sludge. Final rule, 40 CFR Part 503. *Fed. Regist.* **1993**, *58*, 9248–9415.
45. Mathioudakis, V.L.; Kapagiannidis, A.G.; Athanasoulia, E.; Paltzglou, A.D.; Melidis, P.; Aivasidis, A. Sewage Sludge Solar Drying: Experiences from the First Pilot-Scale Application in Greece. *Dry Technol.* **2021**, *31*, 519–526. [CrossRef]

AgriEngineering

MDPI

Technical Note

Testing of Roller-Crimper-and-Undercutting-Blade-Equipped Prototype for Plants Termination

Mino Sportelli, Christian Frasconi, Lorenzo Gagliardi *, Marco Fontanelli, Michele Raffaelli, Massimo Sbrana, Daniele Antichi and Andrea Peruzzi

Department of Agriculture, Food and Environment, University of Pisa, Via del Borghetto 80, 56124 Pisa, Italy
* Correspondence: lorenzo.gagliardi@phd.unipi.it

Abstract: The use of roller crimpers to terminate plants and obtain a natural mulch before cash crop establishment has been identified as a valid and sustainable approach to control weeds. Several enhancements have been evaluated to improve and speed up plant termination to avoid delays in cash crop planting and consequent yield losses, which can occur with standard roller crimpers. In the present study, a new prototype machine provided with a roller crimper and an undercutting blade, allowing it to simultaneously crimp plant stems and cut root systems, has been designed, realized, and tested. The aim of the research was therefore to evaluate the effectiveness of the prototype for plant termination and to compare it with a commercial roller crimper. The termination was performed on a spontaneous vegetation cover (weeds). A monophasic exponential decay model to evaluate the weed termination rate over time was performed. The fitted model showed that the prototype is able to achieve a greater and faster weed devitalization compared to the commercial roller crimper, with a lower plateau (0.23 vs. 5.35 % of greenness of plant material, respectively) and higher constant of decay (1.45 vs. 0.39 day^{-1}, respectively). Further studies are needed to evaluate the prototype's effectiveness in relation to different soil textures, moisture conditions, and amounts of plant biomass to manage, to further improve the machine and extend its use in a broad range of situations, including cover crop termination.

Keywords: mechanical termination; minimum soil tillage; non-chemical methods; cover crops; dead mulch; sustainable agriculture

check for updates

Citation: Sportelli, M.; Frasconi, C.; Gagliardi, L.; Fontanelli, M.; Raffaelli, M.; Sbrana, M.; Antichi, D.; Peruzzi, A. Testing of Roller-Crimper-and-Undercutting-Blade-Equipped Prototype for Plants Termination. *AgriEngineering* **2023**, *5*, 182–192. https://doi.org/10.3390/agriengineering5010013

Academic Editors: Marcello Biocca and Roberto Fanigliulo

Received: 19 December 2022
Revised: 12 January 2023
Accepted: 12 January 2023
Published: 24 January 2023

1. Introduction

Mulching could play a key role in the current agri-food system's transition toward sustainability as a valid physical preventive method for weed control that is useful to reduce reliance on chemical herbicide applications [1–3]. This practice, which consists in covering the soil with organic or synthetic material, prevents weed seeds' germination and emerging seedlings' growth, while favoring water conservation and avoiding temperature fluctuations [4]. Among the available strategies, the termination of cover crops and spontaneous vegetation and the maintenance of their residues as dead mulch on the soil surface (i.e., without disturbing the soil) represents a promising solution that can also improve soil quality by increasing the organic matter content [5,6]. In no-till-based arable and vegetable cropping systems, this practice is usually carried out prior to or simultaneously with cash crop planting, which can take place by means of no-till drills, planters, or transplanters [7–10]. The amount of plant residues plays a key role in terms of weed control effectiveness, as weed emergence tends to decrease as the mulch biomass on the soil's surface increases [11]. The use of the roller crimper to terminate cover crops and obtain a natural mulch has been identified as a valid approach to control weeds that is also able to enhance cash crop yields [12,13]. This device consists of a cylindrical drum with several blades of different shapes placed on the outside. Roller crimpers come in a variety of designs and dimensions, with widths ranging from 2 to 6 m [14–16]. The blades crimp plant stems, cause plant injury, and accelerate senescence, thus terminating them [17,18]. The action of roller crimpers consists in crushing, but not completely cutting, plant stems at equal intervals [17]. Roller crimpers provide intact cover crop residues, ensuring a longer mulch persistence compared to PTO (power take off)-powered mulchers, whose smaller residues degrade more rapidly and are more subject to wind or water displacement [8,19,20]. Roller crimpers also ensures a homogeneous mulch distribution compared to mowing, which together with its long persistence are essential requirements for effective weed control [19,21]. Furthermore, as the cover crop remains anchored to the ground through the roots, there are few probabilities of dragging mulch when the cash crop is planted [16]. Cover crop termination with roller crimpers also showed substantial energetic and agro-ecological advantages compared to the green manure practice commonly used in organic systems, such as higher energy efficiency, higher carbon stored in the soil [22], reduced risk of N leaching [23], and higher predation rate of pests by insects [24].

Roller crimpers ensure a high cover crop termination rate when used from the anthesis of grasses or the 70% flowering stage in forbs [25]. If the cover crop termination is performed before the specific phenological stages, cover crops can survive and compete with the cash crop [25]. A cover crop that has not been killed completely can cause significant yield losses of the cash crop [7,26]. However, in cases of cover crop mixture, it is difficult to find all the species in the appropriate phenological stage for termination at same time [25].

To avoid issues during planting and competition for resources with the cash crop, it is suggested to plant the cash crop after rolling once the residues have reached an appropriate level of desiccation and are easy to penetrate with the equipment [17]. Adapting cash crop planting with the optimal cover crop termination stage may: (i) reduce the time between rolling and planting, causing issues with residues' inadequate desiccation, thus hindering planting [17]; and (ii) delay the cash crop planting with consequent yield losses [27,28]. In the Mediterranean area, timely planting of the spring cash crop is crucial to avoid drought stress issues in the summer period [29]. Several enhancements have been evaluated to improve and speed up cover crops' termination by roller crimpers, even with cover crops at an early phenological stage. To increase their effectiveness, roller crimpers with various designs have been developed, for example with different blade configurations, two-stage roller crimpers, or a water-ballastable sealed roller to provide additional weight [14,30,31]. Herbicide applications have been successfully implemented along with rolling to speed up the cover crop termination in order to allow the planting of the subsequent cash crop at the optimal time [17]. In organic systems where herbicide application is not allowed, Frasconi et al. [5] tested the effect of rolling in combination with flaming. However, the current high operational costs associated with flaming hinder its wider adoption by farmers. Kornecki et al. [32] examined recurrent rolling on cover crops to assess whether biomass termination was accelerated, but this strategy may involve labor management issues and the risk of soil compaction due to multiple passages of the tractor over the same farming land [33]. Antichi et al. [7], on the other hand, evaluated roller-crimping in conjunction with direct drilling of the cash crop, to disturb the cover crop both with the crimping action and the cut performed by the metal disk of the drilling machine. Nevertheless, especially when the intervention was performed on the cover crop at early phenological stages, without the application of glyphosate, plant termination was incomplete, with consequent negative effects on the following cash crop yield.

With the aim of increasing the effectiveness of non-chemical devitalization of both spontaneous vegetation and cover crops, a new prototype machine has recently been developed by the University of Pisa. This machine combines a roller crimper with an undercutting blade working shallowly below the soil surface. This implement, by cutting and therefore damaging the root system, should favor and speed up plant withering, and help to prevent plants regrowth. A machine designed to operate in this way could be very useful, as standard roller-crimpers, acting only on the aboveground portion of plants, cannot prevent in some cases (e.g., in early terminated and vigorous cover crops, or in cover crops stands with high weed infestation) plants' regrowth from the basal buds. To the best of our knowledge, a machine that allows to simultaneously crimp plant stems and cut root systems for an optimal plant termination has never been tested. The aim of the present study is therefore to evaluate the effectiveness of this prototype for the termination of the plant material, in comparison with a commercial roller crimper.

2. Materials and Methods

2.1. The Experimental Layout

The experiment was carried out at the Centre for Agri-environmental Research "Enrico Avanzi", San Piero a Grado (Pisa), Italy (43°40′48″ N, 10°20′49″ E, 1 m.a.s.l., [34]) on a field with homogeneous coverage of spontaneous flora. According to the analysis conducted by the Center for Agri-environmental Research "Enrico Avanzi", the soil was sandy loam (57.07% sand, 22.45% loam, 20.48% clay, 1.92% organic matter, pH = 7.5). From autumn 2019 to spring 2022, the selected field was managed as a permanent grassland through periodic mowing of the vegetation cover. The effectiveness of the prototype in terminating the spontaneous vegetation was evaluated in comparison with a commercial roller crimper on 25 May 2022. The adopted experimental design was a randomized complete block design with three replications.

2.2. Description of the Prototype and the Commercial Roller Crimper

The prototype consists of a three-point-hitch-provided main frame supporting a 1.5 m wide Clemens roller crimper, Eco-Roll 1500 type (Clemens Technologies, Wittlich, Germany); a 1.5 m wide cylindrical drum (Ø 0.20 m) with five vertical cutting discs (each with an external diameter of 0.385 m); and a 1.5 m wide horizontal undercutting blade in hardened steel with a thickness of 8 mm (Figure 1). The cylindrical drum with the cutting discs is positioned between the roller crimper and the undercutting blade, with the function of cutting the crimped vegetable biomass to avoid its accumulation and dragging by the blade (Figure 2). The undercutting blade's working depth can be adjusted by means of telescopic servo rudders, in a range from 3 to 6 cm, according to the characteristics of cover crops and/or spontaneous vegetation. The machine presents a working width of 1.5 m. The mass of the prototype is equal to 610 kg. A hopper has been mounted on the frame to provide additional weight to the prototype to increase the effectiveness of crimping on plant stems when needed. During the experiment, the prototype operated with a working speed of 4 km·h^{-1}, and the working depth of the undercutting blade was set to 4 cm.

Figure 1. Front (**a**) and rear (**b**) view of the prototype: (**1**) roller crimper; (**2**) cylindrical drum with five cutting discs; (**3**) undercutting blade; (**4**) hopper; (**5**) telescopic servo rudders; (**6**) three-point hitch.

Figure 2. Lateral view of the prototype: (**1**) roller crimper; (**2**) cylindrical drum with five cutting discs; (**3**) undercutting blade; (**4**) hopper; (**5**) telescopic servo rudders; (**6**) three-point hitch.

The commercial roller crimper was a Clemens Eco-Roll type with hexagon frame (Figure 3). The machine is provided with two connected roller crimpers, each 1 m wide. The working width can be adjusted hydraulically in the range from 1.10 m to 1.95 m. A tank with maximum capacity of 300 L was attached to the frame, which, when completely filled with water, allows the mass of the machine to increase up to 950 kg. During the experiment, the commercial roller crimper was set to work at 7 km·h^{-1} with a working width of 1.95 m, and the tank was completely filled with water. Both the prototype and the commercial roller crimper were coupled with a Fiat DT 70-90 tractor (FiatAgri, Torino, Italy) powered by a 52.2 kW diesel engine.

(**a**) (**b**)

Figure 3. Front (**a**) and lateral (**b**) view of the commercial roller crimper Clemens Eco-Roll type: ($\mathbf{1_{a,b}}$) roller crimpers; (**2**) hexagon frame; (**3**) tank; (**4**) three-point hitch.

2.3. Data Collection

The spontaneous vegetation was assessed immediately before the termination treatment, both quantitatively and qualitatively. Therefore, weed biomass was measured by cutting and collecting the aboveground biomass present inside a square frame of 0.5 m^2 (1 m long, 0.5 m wide), both in plots subsequently managed with the prototype and in those managed with the commercial roller crimper. Total aboveground fresh biomass was then oven-dried at 100 °C for 3–4 days (until constant weight), and dry biomass was determined. One measurement per replicate of weed biomass before treatment was carried out. Weed flora was characterized by identifying the growth form, growth stage and determining the average height and visual soil cover of the main spontaneous plant species present inside the sampling areas. Two measurements per replicate before treatment were performed for weed height and cover. Average weed height and cover of the experimental field were then determined.

The weed termination rate over time after the machines' intervention was estimated as percentage of greenness of the plant material. The estimate was performed with the analysis of digital images taken on the felled weed [5], using the app Canopeo (Mathworks, Inc., Natick, MA, USA) [35]. The app provides, for each image processed, the percentage of green pixels calculated on the total number of pixels. This method was used because the color of the tissues of a devitalized plant, in which the photosynthetic activity should have been compromised, tends to turn more or less quickly towards colors other than green. Digital images were taken, one for each replicate, inside a 0.5 m^2 square frame, on the day of weed termination (0), and at 1, 2, 3, 4, 5, 6, 8, 10, 12, and 14 days after termination; therefore, on May, 25, 26, 27, 28, 29, 30, and 31, and June 2, 4, 6, and 8, 2022, respectively.

The fuel consumption per hectare related to the machines' performance was estimated on the basis of the theoretical working capacity (which considers that the tractor coupled with the operating machine effectively work at the optimal working speed, and operate for the entire width of action) and the hourly fuel consumption.

The hourly fuel consumption was estimated using the following equation:

$$Ch = W \times d \times Cs \quad (1)$$

where Ch is the hourly fuel consumption of the tractor (kg·h^{-1}), W corresponds to the tractor engine power (kW), d is the effort percentage of the tractor engine required by the operation, and Cs is the tractor engine energetic efficiency (kg fuel·kWh^{-1}). In the present study, based on the tractor characteristics, Cs was estimated at 0.25 kg of fuel·kWh^{-1}, while d was assigned according to the machine type. Values of d equal to 0.35 and 0.55 have been attributed to the commercial roller crimper and the prototype, respectively, considering that the prototype, in addition to rolling, also performs cutting with the vertical cutting discs and the undercutting blade.

2.4. Statistical Analysis

Weed biomass was analyzed with the two-tailed t-test using the statistical software SPSS (IBM Corp, Armonk, NY, USA) to assess any variations in weed infestation before treatment between the plots managed by the two machines in comparison [36]. Weed termination rates over time measured for the plant material felled by the prototype and the commercial roller crimper were analyzed with the statistical software GraphPad Prism version 9.4.1 (GraphPad Software, San Diego, CA, USA) [37]. The software predicted the use of a nonlinear regression by adopting the following monophasic exponential decay model:

$$Y = (Y_0 - Plateau) \times e^{(-kX)} + Plateau \quad (2)$$

where:

- Y corresponds to the dependent variable (i.e., the percentage of green pixels of the felled plant biomass);

- Y_0 is the value of the dependent variable at time 0 (i.e., the percentage of green pixels of the felled plant biomass immediately after the machines' intervention);
- *Plateau* is a parameter estimated by the software and corresponds to the asymptotic value of the dependent variable (i.e., the percentage of green pixels of the felled plant biomass, which occurs at an infinite time);
- X is the independent variable (i.e., the time, which in this case corresponds to days after termination);
- k corresponds to the constant of decay (i.e., a constant estimated by the software that presents as unit the inverse of X (in this case days^{-1})).

3. Results and Discussion

The two-tailed *t*-test revealed no significant differences in terms of weed biomass before treatment between plots where the prototype and the commercial roller crimper were compared, with average values of 337.2 $g{\cdot}m^{-2}$ and 348.6 $g{\cdot}m^{-2}$, respectively. Therefore, it is possible to state that weed biomass before treatment was homogeneous in the experimental field.

The characteristics of the main spontaneous plant species present on the experimental field before treatment, such as growth form, growth stage, average height, and cover, are shown in Table 1.

Table 1. Characteristics of the weed flora present on the experimental field before the treatment.

	Growth Form [1]	Growth Stage (BBCH) [2]	Average Height (m)	Average Cover (%)
Avena sterilis L.	T scap	65	0.70	2.5
Convolvulus arvensis L.	G rhiz	40	0.15	6.2
Lolium multiflorum Lam.	T scap	41	0.40	26.5
Picris echioides L.	T scap	19	0.20	62.0
Verbena officinalis L.	H scap	51	0.25	1.8
Others	-	-	-	1.0

[1] Growth form: T scap—Therophytes scapose; G rhiz—Geophytes rhizomatous; H scap—Hemicryptophytes scapose [38]. [2] Phenological growth stages and BBCH-identification keys of weed species: 65—full flowering: 50% of flowers open, first petals may be fallen; 40—vegetative reproductive organs begin to develop (rhizomes, stolons, tubers, runners, bulbs); 41—flag leaf sheath extending; 19—nine or more true leaves, leaf pairs, or whorls unfolded; 51—inflorescence or flower buds visible [39].

Weed flora identified before the treatment on sampling areas was representative of the Mediterranean area in late spring, and it was mainly composed of species with therophytes scapose growth forms, at different growth stages.

The nonlinear regression conducted on the values of the weed termination rate over time as the percentage of green pixels has preliminarily showed that data relating to the prototype and the commercial roller crimper are better described by two distinct curves (extra sum-of-squares F-test $p < 0.0001$). The parameters shown in Table 2 confirm the goodness of adaptation of the temporal trends of the two machines' weed termination rate to the two different curves, and therefore to the monophasic exponential decay model adopted.

Table 2. Degrees of freedom, coefficient of determination, and coefficient of determination adjusted for the regressions carried out on the temporal trends of weed termination rate of the two machines in comparison.

	Prototype	Commercial Roller Crimper
Degree of freedom	30	30
R^2	0.985	0.981
R^2 adjusted	0.984	0.979

The two monophasic exponential nonlinear regression curves obtained from the percentage of green pixels over time of weeds felled by the prototype and the commercial roller crimper and the main parameters values are shown in Figure 4 and Table 3, respectively.

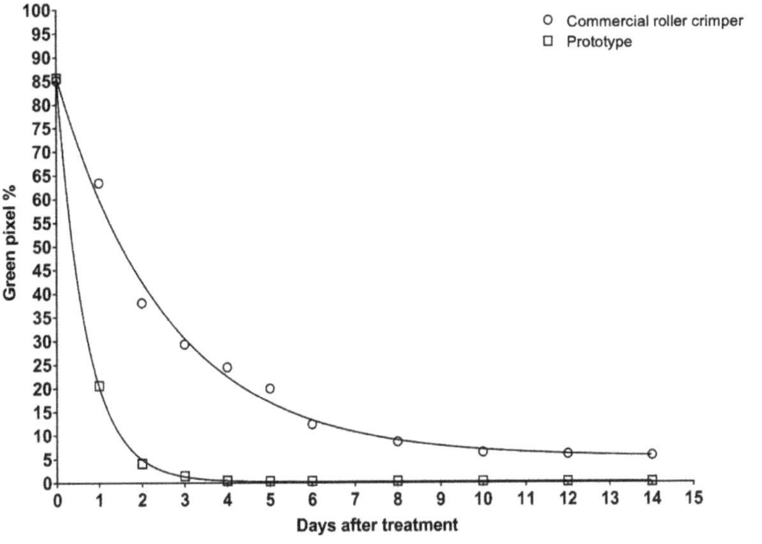

Figure 4. Monophasic exponential nonlinear regression curves of the % of green pixels in the digital images of weed biomass collected at 0, 1, 2, 3, 4, 5, 6, 8, 10, 12, and 14 days after termination.

Table 3. Main parameters of the monophasic exponential nonlinear regression curves.

		Prototype			Commercial Roller Crimper	
Parameters		Ll 95%	Ul 95%		Ll 95%	Ul 95%
Y_0 (%)	85.70	84.76	86.64	85.75	81.82	89.70
Plateau (%)	0.23	−0.11	0.57	5.35	2.73	7.81
k (day^{-1})	1.45	1.40	1.50	0.39	0.34	0.44
Half-life (days)	0.48	0.46	0.49	1.80	1.59	2.04

Y_0 is the percentage of green pixels of the felled weed immediately after the machines' intervention; *plateau* is the percentage of green pixels of the felled weed at an infinite time; k corresponds to the constant of decay; *half-life* is the time (days) required to reach a level of percentage of green pixels equal to 0.5 (Y_0—*plateau*), and it is computed as $\ln(2)/k$; *Ll* corresponds to the lower limit of the confidence interval of the estimate at 95%; *Ul* corresponds to the upper limit of the confidence interval of the estimate at 95%.

From the values of Y_0 and the respective *Ul* and *Ll* of both the tested machines, it is possible to affirm that the prototype and the commercial roller crimper operated under similar conditions of weed cover. This once again supports that the distribution of spontaneous vegetation in the experimental field was rather homogeneous. The *plateau* of the prototype regression curve corresponds to 0.23%. Furthermore, by observing the *Ul* and *Ll* of the parameter's confidence intervals, it is possible to state that the *plateau* of the prototype regression curve does not deviate significantly from the 0% value, with a probability of not less than 95%. This would lead us to hypothesize a greater effectiveness of weed termination of the innovative prototype compared to the commercial roller crimper, whose estimate of the same parameter stands at a value of 5.35%. The higher termination obtained by the prototype could be mainly due to the effectiveness of the undercutting blade. The blade, by separating weed stems from roots, together with the crimping action and the action of the cutting discs, favors a higher withering compared to the commercial roller, which instead only crimps plant stems [40]. However, based on the % of greenness of the felled weeds obtained, it is possible to state that both machines allowed us to reach a plant devitalization above 90%. According to Ashford et al. [18], the achievement of this threshold, in their case with reference to a rye cover crop, is sufficient to plant the cash crop in dried residues without the risk of competition for resources. Concerning the values of the constant of decay *k* and the 95% confidence intervals of the weed termination curves obtained for the machines, the value estimated for the prototype curve is statistically higher, in absolute value, than that of the commercial roller crimper. These first results show that the prototype allows for a faster weed devitalization compared to the commercial roller crimper. The obtained results are also confirmed by the prototype's *half-life* value, which is 3.75 times lower than the commercial roller crimper. The commercial roller crimper's plant termination rate values are consistent with those attained by other authors [41] for a two-stage roller crimper, which led to a plant devitalization of 85% and 96% one and two weeks from the intervention, respectively. Results obtained by the prototype are in agreement with Kornecki and Kichler [42], who found a higher plant termination rate after one and two weeks from the intervention for a flail mower, a machine that performs cutting of plant tissues, compared to roller crimpers. Kornecki [43] observed a significant reduction in the time required to exceed 90% of plant devitalization, similar to that of the prototype, when rolling was performed in combination with chemical herbicide application, and when rolling was carried out three times. However, even if recurrent rolling proved to be a valid alternative to chemical herbicide application to speed up plant termination, multiple passages of the tractor over the same farming land could cause soil compaction [33].

This finding is significant since any delay in plant termination treatment can reduce the time interval between rolling and cash crop planting, making planting difficult if plant residues are not adequately desiccated. Therefore, speeding up plant termination is essential to reduce the risk of delaying the following cash crop planting, and the consequent potential negative effects on yield [17].

Table 4 shows the operative characteristics of the two machines in comparison.

Table 4. Operative characteristics of the prototype and the commercial roller crimper.

		Prototype	Commercial Roller Crimper
Parameters			
Working speed	$km \cdot h^{-1}$	4	7
Working width	m	1.5	1.95
Working capacity	$ha \cdot h^{-1}$	0.6	1.37
Hourly fuel consumption	$kg \cdot h^{-1}$	7.18	4.57
Fuel consumption per hectare	$kg \cdot ha^{-1}$	11.96	3.35

Comparing the working capacity and the fuel consumption of the two machines, it is possible to observe that the prototype presents a lower working capacity, with a decrease of 56.2%, and a greater fuel consumption, with an increase of 257.01%, compared to the commercial roller crimper. The lower working capacity of the prototype is related to its lower forward speed and working width, while the higher fuel consumption is mainly due to the higher effort percentage of the tractor engine required by the prototype operation, which combines rolling and cutting with the vertical discs and the undercutting blade. According to Creamer and Dabney [19], the operation of plant termination by means of a machine that also performs undercutting is slower and may require more power compared to rolling alone. However, as the prototype proved to be more effective at plant termination, repeated rolling or application of flaming, as can occur with ordinary roller crimpers to accelerate plant termination, may not be necessary, thus avoiding the relative disadvantages [32,33]. The lower working capacity of the machine and its higher fuel consumption would make the prototype particularly suitable for more profitable contexts, such as organic horticultural systems. Furthermore, in these contexts, where the creation of raised beds is frequent, the operation of the undercutting blade could also be facilitated [44].

Nevertheless, at this first experimental test, the innovative machine proved to be effective in managing the spontaneous cover, by providing a greater and faster devitalization of weed biomass compared to the commercial roller crimper. Machines that perform plant undercutting, compared to those using other plant termination means, such as mulchers, allow one to create a thicker and longer-lasting mulch, ensuring a greater, season-long weed suppression, and leave a looser soil, facilitating the cash crop planting [44,45]. Furthermore, despite the lower prototype working capacity compared to the commercial roller crimper, the damage to the root system caused by the undercutting blade could contribute to preventing the regrowth of the felled weeds. The optimal and fast termination and the prevention of plant regrowth are crucial to avoid any serious yield losses of the following cash crop [7,26].

4. Conclusions

In the present study, the use of a prototype that simultaneously crimps plant stems and cuts root systems for the management of spontaneous vegetation achieved encouraging results. Indeed, the prototype obtained a higher and faster devitalization of the plant biomass compared to the commercial roller crimper. Therefore, the innovative machine seems to represent a valid tool for the non-chemical termination of spontaneous vegetation or cover crops. A machine with these characteristics can be particularly suitable for horticultural organic systems, allowing the realization of dead mulching, which favors weed control, and other useful ecosystem services such as soil conservation and moisture retention. Further studies are needed to evaluate the prototype's effectiveness in relation to different soil textures, moisture conditions, and amounts of plant biomass to manage, to further improve the innovative machine and extend its use in a broad range of situations. Different shapes, thicknesses, and angles of the undercutting blade should be tested in order to increase the versatility of the prototype in different soil types and moisture conditions. Possible disadvantages have been observed for the prototype's single-termination intervention, such as higher fuel consumption and lower working capacity compared to the commercial roller crimper, as the new machine performs a subsurface cut in addition to rolling. Therefore, it would be desirable to further evaluate the economic aspects and energy consumption within a real farming system. Moreover, it would be useful to investigate the prototype's ability to prevent plant regrowth as a function of the different phenological stage at which plant species are terminated.

Author Contributions: Conceptualization, A.P. and C.F.; methodology, C.F., M.S. (Mino Sportelli), and L.G.; software, C.F.; validation, A.P., C.F., M.F., M.R. and D.A.; formal analysis, C.F. and M.F.; investigation, M.S. (Mino Sportelli), C.F. and L.G.; resources, A.P.; data curation, C.F., M.S. (Mino Sportelli), M.S. (Massimo Sbrana) and L.G.; writing—original draft preparation, L.G., M.S. (Mino Sportelli) and C.F.; writing—review and editing, A.P., C.F., M.F., M.S. (Mino Sportelli) and L.G.; visualization, M.R., M.F., D.A. and M.S. (Massimo Sbrana); supervision, A.P., D.A. and M.R.; project administration, A.P. and C.F.; funding acquisition, A.P. All authors have read and agreed to the published version of the manuscript.

Funding: This research was funded by the Italian Ministry of Agricultural, Forestry and Food Policies (MiPAAF), within the project "Mechanization of organic and conservation horticulture", acronym "MEORBICO", approved by Ministerial Decree 89238 of 19 December 2019.

Institutional Review Board Statement: Not applicable.

Informed Consent Statement: Not applicable.

Data Availability Statement: Not applicable.

Acknowledgments: The authors would like to acknowledge the Agriculture and Environmental Research Centre "Enrico Avanzi" for hosting the trials, and for their technical support; in particular, we would like to thank Alessandro Pannocchia, Giovanni Melai, Marco Della Croce, and Roberta Del Sarto.

Conflicts of Interest: The authors declare no conflict of interest.

References

1. El Bilali, H. Transition Heuristic Frameworks in Research on Agro-Food Sustainability Transitions. *Environ. Dev. Sustain.* **2020**, *22*, 1693–1728. [CrossRef]
2. Monteiro, A.; Santos, S. Sustainable Approach to Weed Management: The Role of Precision Weed Management. *Agronomy* **2022**, *12*, 118. [CrossRef]
3. Hasanuzzaman, M. *Agronomic Crops, Volume 2: Management Practices*, 1st ed.; Springer Nature Singapore Pte Ltd.: Singapore, 2019.
4. Kader, M.A.; Senge, M.; Mojid, M.A.; Ito, K. Recent Advances in Mulching Materials and Methods for Modifying Soil Environment. *Soil Tillage Res.* **2017**, *168*, 155–166. [CrossRef]
5. Frasconi, C.; Martelloni, L.; Antichi, D.; Raffaelli, M.; Fontanelli, M.; Peruzzi, A.; Benincasa, P.; Tosti, G. Combining Roller Crimpers and Flaming for the Termination of Cover Crops in Herbicide-Free No-till Cropping Systems. *PLoS ONE* **2019**, *14*, e0211573. [CrossRef]
6. Warren Raffa, D.; Antichi, D.; Carlesi, S.; Frasconi, C.; Marini, S.; Priori, S.; Bàrberi, P. Groundcover Mulching in Mediterranean Vineyards Improves Soil Chemical, Physical and Biological Health Already in the Short Term. *Agronomy* **2021**, *11*, 787. [CrossRef]
7. Antichi, D.; Carlesi, S.; Mazzoncini, M.; Bàrberi, P. Targeted Timing of Hairy Vetch Cover Crop Termination with Roller Crimper Can Eliminate Glyphosate Requirements in No-till Sunflower. *Agron. Sustain. Dev.* **2022**, *42*, 87. [CrossRef]
8. Mirsky, S.B.; Ryan, M.R.; Curran, W.S.; Teasdale, J.R.; Maul, J.; Spargo, J.T.; Moyer, J.; Grantham, A.M.; Weber, D.; Way, T.R.; et al. Conservation Tillage Issues: Cover Crop-Based Organic Rotational No-till Grain Production in the Mid-Atlantic Region, USA. *Renew. Agric. Food Syst.* **2012**, *27*, 31–40. [CrossRef]
9. Frasconi, C.; Martelloni, L.; Raffaelli, M.; Fontanelli, M.; Abou Chehade, L.; Peruzzi, A.; Antichi, D. A Field Vegetable Transplanter for Use in Both Tilled and No-Till Soils. *Trans. ASABE* **2019**, *62*, 593–602. [CrossRef]
10. Tosti, G.; Benincasa, P.; Farneselli, M.; Guiducci, M.; Onofri, A.; Tei, F. Processing Tomato–Durum Wheat Rotation under Integrated, Organic and Mulch-Based No-Tillage Organic Systems: Yield, N Balance and N Loss. *Agronomy* **2019**, *9*, 718. [CrossRef]
11. Teasdale, J.R.; Mohler, C.L. The Quantitative Relationship between Weed Emergence and the Physical Properties of Mulches. *Weed Sci.* **2000**, *48*, 385–392. [CrossRef]
12. Canali, S.; Campanelli, G.; Ciaccia, C.; Leteo, F.; Testani, E.; Montemurro, F. Conservation Tillage Strategy Based on the Roller Crimper Technology for Weed Control in Mediterranean Vegetable Organic Cropping Systems. *Eur. J. Agron.* **2013**, *50*, 11–18. [CrossRef]
13. Canali, S.; Diacono, M.; Campanelli, G.; Montemurro, F. Organic No-Till with Roller Crimpers: Agro-Ecosystem Services and Applications in Organic Mediterranean Vegetable Productions. *Sustain. Agric. Res.* **2015**, *4*, 70. [CrossRef]
14. Kornecki, T.S.; Price, A.J.; Raper, R.L.; Arriaga, F.J. New Roller Crimper Concepts for Mechanical Termination of Cover Crops in Conservation Agriculture. *Renew. Agric. Food Syst.* **2009**, *24*, 165–173. [CrossRef]
15. Kornecki, T.S.; Price, A.J.; Raper, R.L. Performance of Different Roller Designs in Terminating Rye Cover Crop and Reducing Vibration. *Appl. Eng. Agric.* **2006**, *22*, 633–641. [CrossRef]
16. Vincent-Caboud, L.; Casagrande, M.; David, C.; Ryan, M.R.; Silva, E.M.; Peigne, J. Using Mulch from Cover Crops to Facilitate Organic No-till Soybean and Maize Production. A Review. *Agron. Sustain. Dev.* **2019**, *39*, 45. [CrossRef]
17. Kornecki, T.S.; Price, A.J.; Raper, R.L.; Bergtold, J.S. Effectiveness of Different Herbicide Applicators Mounted on a Roller/Crimper for Accelerated Rye Cover Crop Termination. *Appl. Eng. Agric.* **2009**, *25*, 819–826. [CrossRef]
18. Ashford, D.L.; Reeves, D.W. Use of a Mechanical Roller-Crimper as an Alternative Kill Method for Cover Crops. *Am. J. Altern. Agric.* **2003**, *18*, 37–45. [CrossRef]
19. Creamer, N.G.; Dabney, S.M. Killing Cover Crops Mechanically: Review of Recent Literature and Assessment of New Research Results. *Soil Use Manag.* **2002**, *17*, 32–40. [CrossRef]
20. Liebert, J.A.; DiTommaso, A.; Ryan, M.R. Rolled Mixtures of Barley and Cereal Rye for Weed Suppression in Cover Crop–Based Organic No-Till Planted Soybean. *Weed Sci.* **2017**, *65*, 426–439. [CrossRef]
21. Mirsky, S.B.; Curran, W.S.; Mortensen, D.A.; Ryan, M.R.; Shumway, D.L. Control of Cereal Rye with a Roller/Crimper as Influenced by Cover Crop Phenology. *Agron. J.* **2009**, *101*, 1589–1596. [CrossRef]
22. Diacono, M.; Persiani, A.; Testani, E.; Montemurro, F. Sustainability of Agro-Ecological Practices in Organic Horticulture: Yield, Energy-Use and Carbon Footprint. *Agroecol. Sustain. Food Syst.* **2020**, *44*, 726–746. [CrossRef]
23. Hefner, M.; Gebremikael, M.T.; Canali, S.; Sans Serra, F.X.; Petersen, K.K.; Sorensen, J.N.; De Neve, S.; Labouriau, R.; Kristensen, H.L. Cover Crop Composition Mediates the Constraints and Benefits of Roller-Crimping and Incorporation in Organic White Cabbage Production. *Agric. Ecosyst. Environ.* **2020**, *296*, 106908. [CrossRef]
24. Magagnoli, S.; Masetti, A.; Depalo, L.; Sommaggio, D.; Campanelli, G.; Leteo, F.; Lövei, G.L.; Burgio, G. Cover Crop Termination Techniques Affect Ground Predation within an Organic Vegetable Rotation System: A Test with Artificial Caterpillars. *Biol. Control* **2018**, *117*, 109–114. [CrossRef]
25. Miville, D.; Leroux, G.D. Rolled Winter Rye–Hairy Vetch Cover Crops for Weed Control in No-till Pumpkin. *Weed Technol.* **2018**, *32*, 251–259. [CrossRef]
26. Keene, C.L.; Curran, W.S.; Wallace, J.M.; Ryan, M.R.; Mirsky, S.B.; VanGessel, M.J.; Barbercheck, M.E. Cover Crop Termination Timing Is Critical in Organic Rotational No-Till Systems. *Agron. J.* **2017**, *109*, 272–282. [CrossRef]
27. Carr, P.; Gramig, G.; Liebig, M. Impacts of Organic Zero Tillage Systems on Crops, Weeds, and Soil Quality. *Sustainability* **2013**, *5*, 3172–3201. [CrossRef]

28. Davis, A.S. Cover-Crop Roller–Crimper Contributes to Weed Management in No-Till Soybean. *Weed Sci.* **2010**, *58*, 300–309. [CrossRef]
29. Teasdale, J.R.; Mirsky, S.B.; Spargo, J.T.; Cavigelli, M.A.; Maul, J.E. Reduced-Tillage Organic Corn Production in a Hairy Vetch Cover Crop. *Agron. J.* **2012**, *104*, 621–628. [CrossRef]
30. Kornecki, T.S.; Arriaga, F.J.; Price, A.J. Roller Type and Operating Speed Effects on Rye Termination Rates, Soil Moisture, and Yield of Sweet Corn in a No-till System. *HortScience* **2012**, *47*, 217–223. [CrossRef]
31. Antichi, D.; Tramacere, L.G.; Sbrana, M.; Bendinelli, S.; Frasconi, C. Agronomic Aspects of Cover Crops Termination with Roller Crimper. In Proceedings of the 50th Conference of the Italian Society of Agronomy, Udine, Italy, 15 September 2021; p. 3.
32. Kornecki, T.S.; Kichler, C.M. Influence of Recurrent Rolling/Crimping of a Cereal Rye/Crimson Clover Cover Crop on No-Till Bush Bean Yield. *AgriEngineering* **2022**, *4*, 855–870. [CrossRef]
33. Augustin, K.; Kuhwald, M.; Brunotte, J.; Duttmann, R. Wheel Load and Wheel Pass Frequency as Indicators for Soil Compaction Risk: A Four-Year Analysis of Traffic Intensity at Field Scale. *Geosciences* **2020**, *10*, 292. [CrossRef]
34. *Google Earth Pro Version 7.3.6.9326*; Google LLC: Mountain View, CA, USA, 2022.
35. Patrignani, A.; Ochsner, T.E. Canopeo: A Powerful New Tool for Measuring Fractional Green Canopy Cover. *Agron. J.* **2015**, *107*, 2312–2320. [CrossRef]
36. *Software IBM SPSS*; IBM: Armonk, NY, USA, 2022.
37. *GraphPad Software Version 9.4.1*; Dotmatics: Boston, MA, USA, 2022.
38. Pignatti, S.; Guarino, R.; La Rosa, M. *Flora d'Italia*; Edagricole: Milano, Italy, 2017; Volume 1.
39. Meier, U. *Growth Stages of Mono- and Dicotyledonous Plants (BBCH Monograph)*; Open Agrar Repositorium: Quedlinburg, Germany, 2018.
40. Chicouene, D. Mechanical Destruction of Weeds. A Review. *Agron. Sustain. Dev.* **2007**, *27*, 19–27. [CrossRef]
41. Kornecki, T.S. Effects of Different Rollers and Rye Termination Methods on Soil Moisture and Cotton Production in a No-Till System. *J. Cotton Sci.* **2020**, *24*, 197–210. [CrossRef]
42. Kornecki, T.S.; Kichler, C.M. Effectiveness of Cover Crop Termination Methods on No-Till Cantaloupe. *Agriculture* **2022**, *12*, 66. [CrossRef]
43. Kornecki, T.S. Influence of Recurrent Rolling/Crimping on Cover Crop Termination, Soil Strength and Yield in No-Till Cotton. *AgriEngineering* **2020**, *2*, 631–648. [CrossRef]
44. Creamer, N.G.; Plassman, B.; Bennett, M.A.; Wood, R.K.; Stinner, B.R.; Cardina, J. A Method for Mechanically Killing Cover Crops to Optimize Weed Suppression. *Am. J. Altern. Agric.* **1995**, *10*, 157–162. [CrossRef]
45. Wortman, S.E.; Francis, C.A.; Bernards, M.A.; Blankenship, E.E.; Lindquist, J.L. Mechanical Termination of Diverse Cover Crop Mixtures for Improved Weed Suppression in Organic Cropping Systems. *Weed Sci.* **2013**, *61*, 162–170. [CrossRef]

AgriEngineering

MDPI

Article

Influence of Recurrent Rolling/Crimping of a Cereal Rye/Crimson Clover Cover Crop on No-Till Bush Bean Yield

Ted S. Kornecki * and Corey M. Kichler

USDA-ARS, National Soil Dynamics Laboratory, 411 S Donahue Dr., Auburn, AL 36832, USA
* Correspondence: ted.kornecki@usda.gov; Tel.: +1-334-844-4741 (ext. 2745)

Abstract: A no-till experiment was conducted in Auburn, AL U.S.A. to evaluate the effectiveness of an experimental two-stage roller/crimper in reoccurring rolling over the same area planted with a cereal rye/crimson clover cover crop mix and its influence on bush bean yield. Cover crop termination was much greater with rolling/crimping when compared to the non-rolled (untreated) control. During the three growing seasons, rolling three times had significantly higher termination rates compared to all other treatments, exceeding 90% in 2020. These results suggest that there may be an advantage to rolling/crimping three times so that planting of the cash crop could potentially be performed one week earlier, under favorable soil moisture conditions. However, for growing seasons 2018 and 2020 at three weeks after rolling, there were no differences between rolling treatments. In 2019, rolling three times over the same cover crop area was the only treatment that achieved above 90% termination rate indicating a clear advantage of recurring rolling/crimping in 2019. Rolling/crimping proved to be effective as yield was significantly higher compared to not rolled when averaged over all three growing seasons. This is possible due to the difficulty in planting into a standing cover crop which could have negative effects on seed to soil contact, but more importantly explained with the slight soil moisture advantage given to the rolled plots over the standing cover crop plots. Thus, optimum soil moisture when planting beans is key for successful germination and good main crop stand.

Keywords: roller/crimper; cover crop; no-till drill; conservation tillage

Citation: Kornecki, T.S.; Kichler, C.M. Influence of Recurrent Rolling/Crimping of a Cereal Rye/Crimson Clover Cover Crop on No-Till Bush Bean Yield. *AgriEngineering* **2022**, *4*, 855–870. https://doi.org/10.3390/agriengineering4040055

Academic Editors: Marcello Biocca and Roberto Fanigliulo

Received: 13 July 2022
Accepted: 19 September 2022
Published: 23 September 2022

1. Introduction

The demand for fresh produce grown by small local farms is steadily increasing along with the push for more sustainable no-till agricultural production methods. Cover crop utilization while minimizing inversion tillage has grown in popularity and become a standard practice no-till system for row crop producers. Roller/crimping to kill the cover crop by causing physical injury by crushing plant tissue can be successful if the roller/crimper is aggressive enough to effectively injure the cover crop at the appropriate growth stage [1–4]. The crimping action injures the cover crop by applying a massive vertical pressure to cover crop tissue from the crimping bars against the firm soil surface. The crimping drum with crimping bars equally spaced around the drum's perimeter mechanically injures the plant at equal intervals, leaving a thick layer of residue mulch [2,3]. The crimping effectiveness is directly related to the soil surface firmness and weight of the roller. Soil with higher moisture content is softer which can lead to the crimping bar imprinting the plant into the soil surface instead of crushing it. The advantages of this thick residue layer include retained soil moisture, reduced soil erosion, decreased soil compaction and runoff water, minimized weed seed germination, increased soil organic matter over time, reduced tractor usage and emissions, and carbon sequestration [5–9]. A field experiment conducted in Italy with organically grown zucchini [10] has shown that terminating a barely (*Hordeum distichum* L.) cover crop with a roller/crimper significantly reduced weed pressure (from 6 to 8 times) generating only 770 kg ha^{-1} of weed biomass compared with incorporated cover crop (4840 kg ha^{-1}) or fallow without covers (6020 kg ha^{-1}). However, adoption of these methods is slow for small vegetable farms as there is minimal commercially available equipment options to properly manage cover crop residue, particularly cover crop roller/crimpers that are effective with limited power lighter tractors. Some specific challenges are encountered for organic growers that are not allowed to use commercial pesticides and herbicides in their farm management toolbox [11]. These growers can be overwhelmed with no-till techniques if problems arise such as weed pressure, insects, or disease that must be managed differently than other production systems, such as conventional tillage. For example, with conventional tillage, weeds are often managed with sweep type cultivators to lightly disturb the soil and keep it loose but with no-till methods the soil is covered with desiccated cover crops that are still rooted in the soil making it very difficult to cultivate or hoe weeds. Additionally, no-till using cover crops can increase areas where insects can hide, particularly grasshoppers, that can decimate small transplants [12].

The objective of this experiment was to test the mechanical termination performance of a prototype 1.2 m-wide patented two-stage roller/crimper in a cereal rye and crimson clover cover crop mixture. A bush bean crop was no-till seeded into the rolled residue and pod yield was collected. Without using herbicide, rolling was performed one, two, or three times successively over the same area to see if termination would be accelerated with recurring passes of rolling/crimping operations over the same cover crop area compared to a single pass.

2. Materials and Methods

Cover crops (mixture of cereal rye and crimson clover) were planted in October of each year. Prior to planting the cover crops each year, P_2O_5 fertilizer were applied and incorporated with a rotary tiller at the rates of 65, 20, and 80 kg ha^{-1} on 13 October 2017, 12 October 2018, and 23 October 2019, respectively, according to the soil report for general analysis. Cover crops were planted with a Hoss Garden seeder (Hosstools, Norman Park, GA, USA) with 19 cm row spacing. The planter was calibrated for seeding rates of 50.4 kg ha^{-1} for cereal rye (*Secale cereale*, L., var. Wintergrazer 70) and 14 kg ha^{-1} of pre-inoculated crimson clover seeds (*Trifolium incarnatum*, L., var. Dixie). Rye was planted first and then clover was planted in between each row of rye. Cover crops were terminated between anthesis and early milk growth phase.

A patented 2-stage roller/crimper [13] was designed and specifically built for the Oggun I tractor (CleBer, LLC, Paint Rock, AL, USA) 3-point hitch mid-mount platform (Figure 1). The Oggun I 4-wheel tractor is a power source with a hydrostatic drivetrain (2 rear wheels powered only) by a 16.5 kW Honda GX690 engine and weighs approximately 816 kg (Honda, Tokyo, Japan). The Oggun's mid-mount 3-point hitch feature (Figure 1) can be used for combined operations with another tool mounted on the rear category I, three-point hitch for a single pass. The 2-stage roller has a smooth drum located in the front-most position of the frame (1st stage) and provides stability to the roller frame and serves as the vibration dumper (transferring vibration from the roller's frame into the ground) as it rolls over the cover crop. The crimping drum is constructed from a 11.4 cm (OD) steel tube with 6 pieces of 5.08 cm × 7.62 cm angle iron welded equally spaced on the drum's circumference along its length. Such design provides an aggressive crimping action from the crimping bars, contrary to elliptical (chevron) type rollers that are commercially available. Each of the drums has a 2.54 cm diameter solid steel shaft running through the middle that is supported by compatible pillow block bearings. This crimping drum (2nd stage) is connected with tubular arms that have rubber isolators in the pivot connector and a spring-loaded rod on the opposite end. The drum with crimping bars can pivot independently of the main frame with variable pressure provided from the adjustable spring-loaded rod assembly with a 21 kg cm^{-1} spring rate. For our field testing, the compression spring was preloaded to a distance of 2.54 cm (53 kg force from one spring; 106 kg force from 2 springs) along a crimping bar surface area of 77.4 cm^2, thus applying a static pressure of 1.4 kg cm^{-2} to the cover crop. These springs can be compressed 7.62 cm total. In addition to the force from the springs, the additional downward force comes from the crimping drum assembly weighting 80 kg. Therefore, the total downward force applied to the cover crop is 186 kg every 13.6 cm along the plant's length with downward pressure of 2.4 kg cm^{-2}.

Figure 1. Oggun tractor with mid-mounted patented 2-stage roller/crimper [13]. (**a**) smooth drum for flatting cover crop and serves as the base for roller's stability; (**b**) secondary drum with crimping bars to injure cover crop in equal intervals; (**c**) compression spring to maintain down force for increased crimping efficiency.

Bush beans (*Phaseolus vulgaris*, L., var. Provider) were planted with a Morrison seeder (WHT Foundation, Durham, NC, USA) that was customized to fit on a 3-point hitch (Figure 2). The Morrison seeder is a single row planter unit originally designed for a two-wheel walk-behind tractor to plant a cash crop in no-till systems. This planter was also modified to fit a patented variable depth cutting coulter (Figure 2b) that is powered by a hydraulic motor and roller chain drive with the depth controlled with an electric linear actuator [14]. The variable depth cutting coulter system was designed to improve cutting of heavy cover crop residue for small scale planters where power and weight of the implement would limit cutting effectiveness compared to larger machines.

Figure 2. Modified Morrison no-till drill with powered coulter for effective cover crop cutting. (**a**) rubber shield to press cover crop against the soil; (**b**) powered coulter with hydraulic motor drive to cut the cover crop residue and topsoil [14]; (**c**) secondary coulter with welded spikes (on opposite side: not shown) for better engagement with the soil and to power the metering unit of the drill; (**d**) electric linear actuator to control the depth of the powered coulter in the soil; (**e**) metering unit assembly with seed discharge tube, seed dispensing box, and rubber closing wheel.

The experiment was conducted at the National Soil Dynamics Laboratory in Auburn, AL, USA, (32.61° N, −85.48° E) on a Davidson Clay soil having 25% sand, 31% silt, 44% clay (a clayey kaolinitic thermic (oxidic) Rhodic Paleudults). The experiment started with planting cover crops in October of 2017 and was concluded in July of 2020 for a total of 3 complete growing cycles (seasons). Rolling treatments were applied according to the plot layout with standing plots used as a control. The experimental layout, depicted in Figure 3, consisted of four different treatments in a randomized complete block design (RCBD) configuration. The four treatments included R1 (rolled once), R2 (rolled twice), R3 (rolled three times) over the same cover crop area, with the control (C) for comparison (standing control: untreated. Treatments were randomized within each block. All rolling/crimping treatments were completed in the same day. Due to space constraints, the standing plots were 4.57 m and the rolled plots were 6.1 m. Four border plots were included to allow space for equipment maneuvering with a length of 3.05 m.

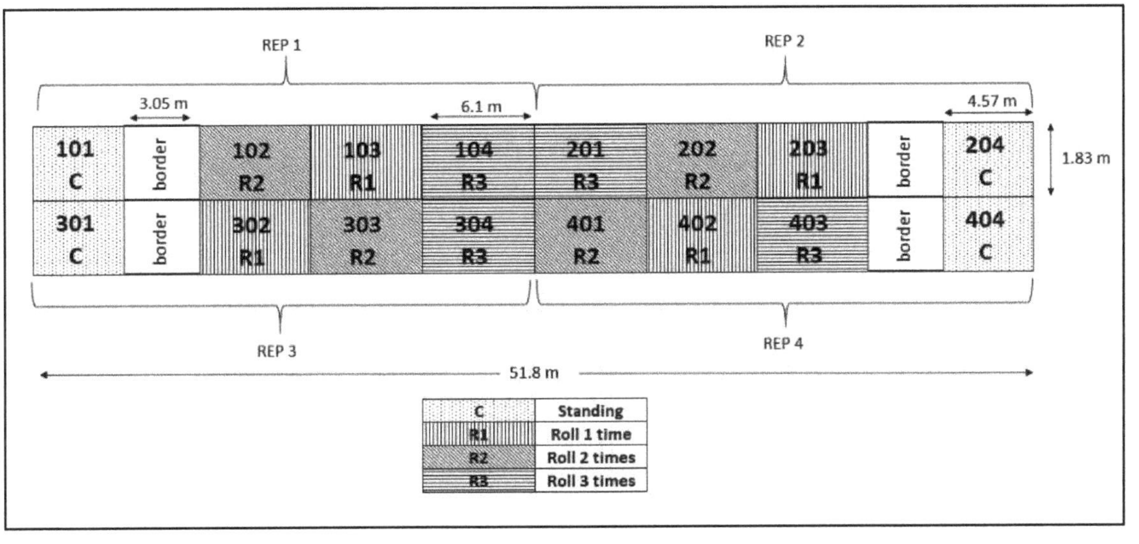

Figure 3. Experiment layout: the randomized block design with four replications.

Immediately prior to applying termination treatments, cover crop production data was collected including biomass and plant heights. A single 0.25 m^2 biomass sample was collected per plot along with 6 heights for each cover crop per plot (i.e., 6 per rye; 6 per clover for each plot). Biomass samples were cut, placed in paper bags, then the samples were placed in a programmable electric shelf oven with forced air flow by convection for 24 h at a temperature of 55 °C (Model No. SC-400 manufactured by Grieve Corporation, Round Lake, IL, USA) to dry down and remove water content from the sample. After the drying process, the cover crop samples were then weighed and recorded. Plant heights were collected using a foldable measuring stick from the soil surface to the top of the seed head of both rye and clover.

Termination data were collected utilizing the SPAD 502 chlorophyll meter (Spectrum Technologies, Aurora, IL, USA). Since cover crop species were not separated for individual biomass data, it was assumed that rye accounted for 80% of the plot cropping area and the clover accounted for the other 20% of the plot cropping area. These percentages were used to give weighted termination values by crop to the termination data collected with the SPAD chlorophyll meter. This was a way to give more weight to the rye compared to the clover regarding percentage kill data (termination) which is more representative of each of the crop's contribution to the mixture. To evaluate the cereal rye and crimson clover termination rates, data collected with a handheld SPAD chlorophyll meter was converted utilizing a linear regression equation and procedure described by Kornecki et al. (2012) [15]. Volumetric soil moisture content (VMC) using the time domain reflectometry soil moisture meter TDR300 (Spectrum Technologies, Aurora, IL, USA). All data were collected weekly for 3 weeks after the termination was complete. Plant chlorophyll content data from 0 to 50 scale, where 0 is 100% of termination (no chlorophyll activity) and 50 is 0% termination rates (plant green with full chlorophyll activity) were collected 3 times per plot with individual leaf samples of each species (3 per species per plot) and VMC was collected 3 times per plot.

After week 3, a single row of bush beans was planted into each plot using a Morrison planter (Figure 3) with the patented variable depth cutting coulter system [14]. Successive harvests were collected approximately two times per week depending on plant production. In 2018 and 2020, there were 6 bean harvests, whereas 7 harvests occurred in 2019. The harvested beans were then weighed, and the weight was recorded by plot. The field activities during three growing seasons are presented in Table 1.

Table 1. Detailed field activities of the experiment during three growing seasons 2018–2020.

Field Activities	Growing Season		
	2018	2019	2020
Cover crop planting	20 October 2017	12 October 2018	24 October 2019
Collect biomass and plant heights	17 April 2018	15 April 2019	3 April 2020
Collect mc, chlor-wk0	18 April 2018	16 April 2019	6 April 2020
Termination treatment applied	18 April 2018	16 April 2019	6 April 2020
Week 1 termination data	25 April 2018	24 April 2019	14 April 2020
Week 2 termination data	2 May 2018	30 April 2019	20 April 2020
Week 3 termination data	9 May 2018	7 May 2019	27 April 2020
Planted bush beans	7 June 2018	16 May 2019	22 May 2020
Harvested beans #1	20 July 2018	2 July 2019	8 July 2020
Harvested beans #2	25 July 2018	8 July 2019	14 July 2020
Harvested beans #3	30 July 2018	12 July 2019	17 July 2020
Harvested beans #4	3 August 2018	18 July 2019	22 July 2020
Harvested beans #5	8 August 2018	24 July 2019	28 July 2020
Harvested beans #6	17 August 2018	31 July 2019	31 July 2020
Harvested beans #7	-	6 August 2019	-

Weather data (AWIS, 2021) [16] are presented in Table 2 which show cumulative precipitation and the average ambient minimum and maximum temperatures for specific periods of agronomic activities during growing seasons (from 2017 to 2020) which had an influence on cover crop production and bush bean yields.

Cover crop plant length and biomass, termination data, volumetric soil moisture content, and bean yield were subjected to analysis of variance and treatment means were separated using the Fisher's protected Least Significant Differences (LSD) test at the 0.10 (10%) probability level. Cover crop mixture and roller/crimper were considered fixed effects and years were considered random effects [17]. Where interactions between treatments and weeks or years occurred, data were analyzed separately and where no interactions were present, data were combined using SAS [18], ANOVA Analyst's linear model.

Table 2. Rainfall amounts and ambient temperatures (minimum and maximum) in each growing season during specified periods of agronomic field activities.

Time Period for Specific Agronomic Field Activities	Cumulative Rainfall (mm)			Average Minimum Temperature (°C)			Average Maximum Temperature (°C)		
	2018	2019	2020	2018	2019	2020	2018	2019	2020
Time within one month before cover crop planting *	126	160	53	17.7	21.9	18.0	27.9	30.9	29.2
Time within one week before cover crop planting *	0	101	30	14.5	21.7	11.2	25.5	29.6	21.9
Time between planting cover crop and its termination *	601	670	861	5.0	7.7	7.5	17.1	18.2	18.3
Time within one month before cover crop termination	212	54	28	6.6	10.1	12.8	22.4	21.8	24.0
Time within one week before cover crop termination	100	29	5	4.8	16.2	10.3	23.3	27.0	24.1
Time within three weeks of termination	37	80	224	12.1	14.1	12.1	24.7	25.9	23.5
Time from third week of termination to beans planting	96	73	29	20.1	17.4	13.6	29.5	27.1	25.3
Time from planting cash crop to its first harvest	754	82	157	20.7	20.9	20.6	31.5	31.3	30.0
Time between first harvest and last harvest	59	92	86	22.1	22.1	22.8	31.3	32.4	32.3

* Planting of cover crops was accomplished in preceding fall periods (e.g., cover crop for 2018 growing season was planted in October of 2017).

3. Results and Discussion

3.1. Plant Length and Biomass Production for Cereal Rye and Crimson Clover Mixture

Plant lengths for both variables: CROP (cover crop) and YEAR (growing season) were highly significant with their respective $p < 0.0001$ and 0.0004. In addition, there were significant interactions between CROP and YEAR variables with $p < 0.0001$ (Table 3), so analysis was performed separately for each cover crop and each year. Results for the plant length are presented in Table 4. For cereal rye, the length was significantly greater for the 2020 growing season reaching 166.1 cm followed by 151.1 cm in 2019 and 159.6 cm in 2018. The crimson clover had a significantly greater length of 74.1 cm in 2018, followed by 67 cm in 2019, and the shortest length of 56.8 cm was observed in 2020. The length of the clover decreased consecutively every year which was most likely explained by its declining stand establishment and contribution to the total biomass. However, the biomass samples were not separated by species, as the total combined weight of cover crop mixture was assessed. Across all growing seasons, average plant length was 158.9 cm and 66 cm for cereal rye and crimson clover, respectively. Results from on-farm replicated field experiment in central Alabama [12], have shown that the length for cereal rye and crimson clover in mixture was 149.4 and 54.4 cm, respectively. Similar results for crimson clover were found in previous research [19] at northern Alabama generating average plant height of 165.4 cm for cereal rye and 54.1 cm for crimson clover.

Table 3. ANOVA results with respect to cover crop plant length and biomass mixture.

Cereal Rye and Crimson Clover Length			Cover Crop Mixture Biomass		
Effect	F-Value	*p*-Value	Effect	F-Value	*p*-Value
REP	3.19	0.0276	REP	1.08	0.3721
YEAR	8.67	0.0004	YEAR	33.21	<0.0001
CROP	3524.65	<0.0001	TRT	0.25	0.8620
CROP*YEAR	27.27	<0.0001	YEAR*TRT	0.64	0.7000

Table 4. Plant length in the mixture of cereal rye/crimson clover and combined biomass mixture.

YEAR	Cereal Rye Length (cm)	Crimson Clover Length (cm)	Mixture Biomass kg ha^{-1}
2018	159.6 b *	74.1 a	11,646 a
2019	151.1 c	67.1 b	7398 b
2020	166.1 a	56.8 c	7659 b
p-value	0.0002	<0.0001	<0.0001
LSD	5.4	3.5	989

* Same lower-case letters indicate no yield difference in each column among growing seasons.

There were significant differences in biomass amounts of cover crop mix among the years ($p < 0.0001$, Table 3) indicating that different weather and climatic conditions affected the biomass production presented in Table 4, especially the biomass of clover in the mix. The first year, 2018, produced the highest amount of biomass measured at 11,646 kg ha^{-1} and significantly more than other years. Both 2019 and 2020 showed statistically similar biomass amounts with 7398 and 7659 kg ha^{-1}, respectively, but both were significantly lower than that produced in 2018. These results represent average biomass production in Alabama cited in previous study [19–21], and these biomass amounts were weather related.

During the three-year field experiment conducted on-farm in central Alabama with different cover crops [12], the dry biomass of cereal rye and crimson clover mixture ranged between 4712 and 8120 kg ha^{-1} with average biomass of 6965 kg ha^{-1} across 2009 to 2011 growing seasons. In another experiment [22] conducted in Indiana at three sites and 2 growing seasons, similar biomass production of cereal rye and crimson clover mixture was between 5451 and 8144 kg ha^{-1}, with average biomass of 6703 kg ha^{-1} across years and sites. According to researchers [23] who conducted a multiyear field experiment in North Carolina, USA, the biomass of cereal rye and crimson clover mixture was between 3820 and 6610 kg ha^{-1}, but they stated that these levels were below expectation of >8000 kg ha^{-1} to provide adequate weed control. In fact, under optimum weather conditions and fertilization, researchers [24] reported that in that region, cereal biomass can exceed 9000 kg ha^{-1}, whereas biomass for crimson clover can reach 5500 kg ha^{-1}. Despite some similarities in biomass production, the range of differences in growing seasons and locations were mainly dependent on different weather and soil moisture conditions.

The higher biomass for the 2018 season can be explained by the higher soil available moisture during the most vigorous cover crop growth in the spring (March and April). As showed in Table 2, during one month before cover crop termination the highest rainfall amount of 212 mm was reported in 2018 growing season compared with 54 mm in 2019 (74.5% less than in 2018) and the lowest rainfall of 28 mm in 2020 (86.8% less than in 2018). These similar rainfall trends continued one week before applying rolling treatments to terminate cover crop mixture. In 2018, one week before cover crop termination rainfall amount was 100 mm compared to lower amounts of 29 mm in 2019 and only 5 mm in 2020. Biomass results for the cereal rye during the three growing seasons do not correspond with cereal rye heights as the tallest stems do not produce higher biomass, i.e., shorter plants can be thicker and might generate larger biomass amounts. In contrast, these rainfall amounts impacted crimson clover heights with the tallest plant of 74 cm in 2018 indicating greater biomass production, compared to 67 cm and 56 cm for 2019 and 2020, respectively, generating lower biomass.

3.2. Cover Crop Mixture Termination

The rate at which the cover crop died was assessed with a chlorophyll meter. This meter was used to collect the chlorophyll activity on three individual leaves of each cover crop (rye and clover) to obtain an accurate assessment. The initial analysis of variance results presented in Table 5 indicates that significant differences in cover crop termination were reported for variable YEAR, WEEK, and TRT all with $p < 0.0001$. In addition, the interactions between WEEK*TRT variables were also significant ($p < 0.0001$). Therefore, the analysis of variance was performed separately by YEAR and then presented by week for each rolling treatment.

Table 5. Analysis of variance with respect to cover crop mixture termination.

Effect	F-Value	*p*-Value
REP	2.49	0.0620
YEAR	30.47	<0.0001
WEEK	1102.42	<0.0001
TRT	554.76	<0.0001
YEAR*TRT	1.71	0.1210
YEAR*WEEK	1.72	0.1189
WEEK*TRT	66.93	<0.0001

3.2.1. 2018 Termination Rates

For 2018, week 0 termination ranged from 9.7% (standing) to 15.9% (rolled once) (Table 6). For week 1, all rolling treatments were statistically similar with an average of 56.8% compared to a significantly lower Standing amount of 13%, illustrating that the rolling treatments were very effective at advancing the termination of the cover crops at one week after rolling. However, 56.8% termination rate is not sufficient for planting the subsequent cash crop, as it is recommended that planting of the main cash crop be done at rates of 90% or greater [1]. The results for week two after rolling showed significant difference for the rolling three times treatment with the highest reported termination rate of 86.8% followed by 73.3% for rolling twice and 63.7 % for rolling once. Rolling multiple times causes more injury with every pass and is illustrated with the termination rates being in a step sequence. The standing control was significantly less compared to all rolling treatments at 18.8%. The rolling three times treatment at week 3 was the only treatment to achieve greater than 90% termination rate measured at 91.2%. Similar results were obtained by [25] when in one growing season, termination rates for mixture (cereal rye, crimson clover and hairy vetch) rolling three times with two-stage roller/crimper exceeded 95%. However, rolling twice and rolling once treatments were not different statistically with 87.6% and 88.8%, respectively, indicating that these where at a high enough death percentage for adequate planting conditions without competition between the cash crop and cover crop for moisture and nutrients.

Table 6. Cover crop termination results (%) during 2018, 2019, 2020 growing seasons for cereal rye/crimson clover mixture.

Rolling Treatment	Week 0	Week 1	Week 2	Week 3
		2018		
Standing	9.7 b *	13.0 b	18.8 d	45.5 b
Rolling once	16.0 a	58.3 a	63.7 c	88.8 a
Rolling twice	14.6 a	55.7 a	73.3 b	87.6 a
Rolling three times	12.7 ab	56.5 a	86.8 a	91.2 a
p-value	0.0776	<0.0001	<0.0001	<0.0001
LSD	3.94	11.13	7.20	5.82
		2019		
Standing	9.3	12.9 c *	14.5 c	28.7 d
Rolling once	10.4	45.0 b	69.9 b	75.5 c
Rolling twice	5.6	49.1 ab	72.5 b	83.6 b
Rolling three times	4.7	54.8 ab	80.4 a	91.1 a
p-value	0.3924	<0.0001	<0.0001	<0.0001
LSD	N/S	9.10	5.06	4.15
		2020		
Standing	12.1	21.6 c *	16.6 d	31.4 b
Rolling once	13.6	51.9 b	68.2 c	96.5 a
Rolling twice	14.4	58.5 b	79.1 b	94.7 a
Rolling three times	13.8	70.9 a	91.7 a	97.9 a
p-value	0.7701	<0.0001	<0.0001	<0.0001
LSD	N/S	7.88	5.05	7.00

* Same lower-case letters indicate no yield difference in each column at each week of the evaluation.

3.2.2. 2019 Termination Rates

Termination data by treatment and week are presented in Table 6. For 2019, week 0 termination, no significant differences were found among all rolling treatments and the control ranging from 4.7% to 10.4%. These numerical values were associated with the slight differences in cover crop maturity within the experimental area, since at week 0, termination data were collected before rolling treatment application. At week 1 after rolling, a significantly higher termination rate was obtained for rolling three times at 54.8 % compared to rolling once (45.0%) but rolling twice at 49% was not significantly different than rolling three times. The control was significantly lowest for week one after rolling at 12.9%. The rolling three times treatment for week two was significantly highest at 80.4%. The once and twice rolled termination rates were similar at 69.9% and 72.5%, respectively, with the control being the lowest at 14.5%. At week three after rolling, termination rates for all rolling treatments were significantly different. The rolling three times treatment has the highest termination rate at 91.1%. These results follow findings from previous field experiment conducted in northern Alabama, USA with cover crop mixture (cereal rye crimson clover and hairy vetch) [25] generating 97% termination rates rolling three times at three weeks after rolling. The rolling twice treatment was lower than rolling three times at 83.6% followed by rolling once having termination rate of 75.5%. The lowest termination rate was associated with the control at only 28.7%.

3.2.3. 2020 Termination Rates

For the 2020 season at week 0, no differences existed between any of the treatments and the control (Table 6). Week 1 showed an advantage in the rolling three times treatment at 70.9%, which is the highest kill rate for week 1 for all years. The once and twice treatments were similar at 51.9% and 58.5%, respectively. The untreated control had the lowest termination rate at 21.6%. The rolling three times treatment generated a termination rate of 91.7%, already at the second week after rolling, again having the highest kill rate for week 2 out of all years and treatments. The rolling two times treatment was second best at

79.1% followed by rolling once at 68.2% compared to the control at 16.6%, all significantly different. For week 3, the rolling treatments measure similarly with an average of 96.4% compared to the lower control at only 31.4%. For 2020, rolling three times showed a significant advantage over only rolling once or twice providing a termination rate of 91.7% at 2 weeks after rolling signifying that the cash crops could be planted earlier compared to the other treatments.

Overall, results from all growing seasons indicate that cover crop termination rates for week two or three after rolling illustrates the advantage of rolling three times by generating termination rates exceeding 90% that according to Ashford and Reeves (2003) [1] are sufficient rates to plant a cash crop into desiccated residue cover. According to [21] rolling cereal rye three times with a two-stage roller/crimper generated termination rates consistently above 90% (91–100%) after 7, 14, and 21 days, indicating that planting of main crop can be accomplished earlier than three weeks after cover crop termination. These results also agreed with other studies [26–28] showing that three times rolling over the same cover crop area accelerates termination rates using rollers/crimpers which is very important in organic no till-systems with cover crops, as using commercial herbicides is prohibited and efficient termination of cover crops is solely dependent on mechanical termination by rollers/crimpers. The advantages of earlier cash crop planting can be a reduced weed, insect, and disease pressure compared to later planting.

3.3. Soil Volumetric Moisture Content (VMC)

Volumetric moisture readings were collected weekly starting at week 0 (day of termination) up to week 3 (3 weeks after termination) to compare the soil moisture amounts between the standing control plots and the rolling treatments. The initial results (Table 7) show difference for YEAR, WEEK, and TRT with $p < 0.0001$. The interactions of YEAR*WEEK and WEEK*TRT were also significant ($p < 0.0001$), therefore, the statistical analysis for this experiment was separated by YEAR and then presented by week for all treatments.

Table 7. Analysis of variance for the soil VMC.

Variable	F-Value	*p*-Value
REP	11.27	<0.0001
YEAR	131.75	<0.0001
WEEK	110.30	<0.0001
TRT	51.01	<0.0001
YEAR*TRT	0.71	0.6424
YEAR*WEEK	131.67	<0.0001
WEEK*TRT	4.34	<0.0001

3.3.1. Soil VMC in 2018 Growing Season

VMC results from 2018 growing season are shown in Table 8. The VMC for week 0 showed no significant difference between any of the treatments and the control, which was expected as data were collected before rolling treatment application. For week one, significantly higher VMC of 15.5 % was observed for rolling three times (R3) and 16.5% for rolling twice (R2) without significant difference between rolling three times treatment (R3) and rolling once treatment (R1) having VMC of 14.8%. The lowest volumetric soil moisture content was measured for the standing cover crop at 11.6%, which is realistic, considering that the standing cover crop is still actively growing, and the soil surface is not completely covered, allowing more moisture loss to the air (i.e., more soil evaporation). For week 2, the rolling treatments were all statistically similar ranging from 12.9% (R3) down to 11.89% (R1). The lowest for week 2, was again the standing control treatment at 9.4%. For week 3, the results experienced significant but decreased separation with a $p = 0.0702$, which shows that the cover crops, including the standing, are consuming less soil moisture for growth as plants mature and its termination rates have advanced. The R1, R2, and R3 treatments were statistically similar with the R2 treatment also being similar to the untreated control (standing) cover crop. Similar results were obtained by researchers [3] who examined rolling/crimping effect of different rollers/crimpers on volumetric soil moisture content. Utilizing mechanical termination and comparing with standing cover rye crop, an average VMC in 2006 was 10.5% using two-stage roller/crimper vs./7.1% for standing rye cover crop. In 2007, when a severe drought occurred during the evaluation period of 3 weeks after rolling, VMC with two-stage roller was 3.3% compared to 1.8% VMC for untreated cover crop. In 2008, the soil VMC with two-stage roller was 6.9% compared with 4.1% for an untreated (standing) rye cover crop. These results clearly indicate that covering soil surface with flattened and terminated cover crop residue conserves soil water.

Table 8. Soil volumetric moisture content (%) in 2018, 2019, 2020 growing season assessed from rolling treatment application up to three weeks after rolling.

Rolling Treatment	Week 0	Week 1	Week 2	Week 3
	2018			
Standing	13.0	11.6 c *	9.4 b	8.3 b
Rolling once	13.4	14.8 b	11.8 a	9.6 a
Rolling twice	13.2	16.5 a	12.1 a	9.4 ab
Rolling three times	13.4	15.5 ab	12.9 a	10.4 a
p-value	0.3848	<0.0001	0.0071	0.0702
LSD	N/S	1.06	1.38	1.21
	2019			
Standing	11.6 b *	11.9 b	11.7 b	14.1 b
Rolling once	12.5 ab	14.7 a	13.1 ab	16.5 a
Rolling twice	11.7 b	14.9 a	13.7 a	17.1 a
Rolling three times	13.2 a	15.7 a	14.5 a	17.1 a
p-value	0.0840	0.0261	0.0349	0.0017
LSD	1.12	1.93	1.48	1.08
	2020			
Standing	9.7	14.0 b *	19.3 b	11.4 b
Rolling once	9.4	17.6 a	21.7 a	14.3 a
Rolling twice	10.7	18.9 a	22.0 a	14.2 a
Rolling three times	10.5	18.1 a	21.7 a	14.3 a
p-value	0.2026	0.0074	0.0738	0.0007
LSD	N/S	2.00	1.77	0.94

* Same lower-case letters indicate no soil VMC difference in each column at each week of the evaluation.

3.3.2. Soil VMC in 2019 Growing Season

In contrast to 2018, the 2019 growing season noticed statistical differences for week 0 even though the treatments were not applied yet which could be contributed to in-field variability (Table 8). For week 1, the rolled treatments showed no statistical differences, however VMC numerical values were greater compared to the standing control plots. Similar results were detected for weeks 2 and 3, however for week 2, similarities were observed for the rolling once treatment compared to the standing control plots. A deeper look into week 3 shows the average VMC for the rolling treatments to be 16.9% which is 20% more soil moisture compared to the standing plots. VMC results by treatment within each week seemed to follow a trend for 2019 in which the VMC was numerically higher for rolling three times and then decreasing order according to the number of roll/crimp passes. Faster termination occurred with the 3X rolling which noticed an increase in VMC compared to the other treatments, although it was not significantly different than the rolling twice or rolling once treatment. The standing control plot was significantly less than all the rolling treatments for weeks 1, 2, and 3 while the cover crop is still actively growing and consuming soil moisture to develop compared to the rolled/crimped treatments. Results from another field experiment in Northern Alabama conducted in 2011 [28] supports these findings, as the VMCs for rolled/crimped rye residue by two-stage roller/crimper were significantly higher: 18.2%, 13.3% and 19.0% compared with the untreated control of 12.6%, 6.6% and 11.9%, at 1, 2, and 3 weeks after rolling, respectively.

3.3.3. Soil VMC in 2020 Growing Season

In 2020, no treatment differences were observed for week 0 (Table 8). The same trend existed for weeks 1, 2, and 3 with all rolling treatments being similar within weeks but statistically different than the standing cover control treatment. Rainfall events did occur between week 0 and 1 readings along with a rainfall event in the amount of 144 mm [16] occurring on the day when the week 2 VMC readings were collected (Table 2) which explains the increase in the average VMC at week 2 for rolled/crimped treatments of 21.8% compared to 19.3 for the standing (control). Similar results with increased VMC were reported by researchers [28] where in 2010 three weeks after rolling, VMC for rolled/crimped rye residue by the two-stage roller/crimper was 26.0% compared to 21.9% for the control after two rainfall events with the total rainfall amount of 43 mm.

Overall, volumetric soil moisture results obtained during three growing seasons, consistently showed that rolling down cover crops against soil surface conserves soil water. In contrast, for an untreated cover crop mixture (standing cereal rye and crimson clover), there was more bare soil exposed between plants, allowing for more soil evaporation. In addition, evapotranspiration of still-living rye and clover plants further depleted soil moisture, thus not conserving soil water. These findings agreed with several previous field studies with cover crops [2,3,5–7,27] in which benefits from cover crops residues were identified, such as increased water holding capacity due to a mulch effect.

3.4. Bush Bean Yield

Based on ANOVA results (Table 9) from three growing seasons, there was significant difference in the bush bean yield with respect to YEAR ($p = 0.0002$) and TRT ($p = 0.0668$) variables. However, there were no significant interaction between YEAR and TRT, therefore differences for main effects (YEAR and TRT) are analyzed separately and reported by year with respect to rolling treatments.

Table 9. Analysis of variance results for bush beans yield.

Variable	F-Value	*p*-Value
REP	1.02	0.3948
YEAR	11.56	0.0002
TRT	2.63	0.0668
YEAR*TRT	0.72	0.6371

The bush bean yield for the 2018 season averaged over all rolling treatments was significantly higher producing 23,160 kg ha^{-1} when compared to lower yield of 19,892 kg ha^{-1} in 2020 (14.1% lower than in 2018) and the lowest yield of 16,838 kg ha^{-1} (27.3% lower than in 2018) was obtained in 2019 (Table 10). The main reason for the yield difference in each growing season was the amount of available water to grow plants. In fact, the total rainfall amount in 2018 from planting the beans to their first harvest was 754 mm, compared to a much lower rainfall of 157 mm (20.8 % of 2018 rainfall) in 2020 and the lowest rainfall amount of 82 mm (10.2% of 2018 rainfall) that was received in 2019 (Table 2). As shown in Table 11, for both growing seasons of 2018 and 2019, no significant differences in the yield were observed among rolling treatments and the control. For 2020, the rolling treatments showed statistically higher yield with an average of 21,153 kg ha^{-1} compared to the lower yield of the control at 16,109 kg ha^{-1}.

Table 10. Bush bean yield (kg ha^{-1}) at each growing season averaged over rolling treatments.

YEAR	Bush Bean Yield
2018	23,160 a *
2019	16,837 c
2020	19,892 b
p-value	0.0002
LSD	2225.5

* Same lower-case letters indicate no yield difference in the second column among growing seasons.

Table 11. Bush beans yield (kg ha^{-1}) in each growing season and average yield over all growing seasons with respect to rolling treatments.

Rolling Treatment	2018	2019	2020	Average over Years by Treatment
Standing (untreated)	23,126	13,032	16,109 b *	17,422 b
Rolling once	21,718	18,084	21,013 a	20,272 a
Rolling twice	23,751	17,647	21,430 a	20,943 a
Rolling three times	24,044	18,587	21,015 a	21,215 a
p-value	0.4551	0.2777	0.0270	0.0668
LSD	N/S	N/S	2951.9	2570.0

* Same lower-case letters indicate no yield difference in each column among rolling treatments.

Across treatments and years, the overall average pod yield was 19,963 kg ha^{-1}. Similar results were obtained from a field experiment conducted by a researcher [29] in Oregon, USA, who examined water availability effect on bush beans, and reported pod yields between 15,864 kg ha^{-1} and 19,348 kg ha^{-1}. In another field experiment conducted [30] in India with different biostimulants, the total produced pod yield was 12,600 kg ha^{-1}, which was about 37% lower than yield obtained from this study and [29]. A yield of main crops under organic no-till farming with cover crops is also very dependent of the geographical location. In fact, a significant cabbage yield reduction (68–100%) with cover crops terminated by a roller/crimper was reported by European researchers [31] from a multi-location study in Denmark, Estonia and at three locations in Belgium (northern and western Europe). Results from [31] indicated that the main reason in seven out of nine cases was mainly due to slower mineralization/degradation of cover crop residues and reduced soil mineral nitrogen availability. In contrast, rapid cover crop degradation in Alabama's subtropical climate with higher temperatures allow to release soil nitrogen that is available to main crops.

On average over all growing seasons, the rolled three times treatment had slightly higher numerical value for yield (21,215 kg ha^{-1}) when compared to other treatments, but these numerical values were not statistically different. This is most likely due to the increased cover crop death rate that allowed for slightly better planting conditions including soil moisture and cover crop plant brittleness. However, averaged across all years, rolling treatments yielded more beans compared to the standing treatments. This emphasizes the importance of cover crop management using roller/crimpers to retain soil moisture for better bush bean establishment.

3.5. Economic Considerations

In the middle of 2022, the United States national average price for regular grade gasoline is US $1.04 L^{-1} [32] which is higher due to international tensions and inflation. The Oggun tractor having a hydrostatic drive, will be operated at full throttle with a fuel consumption rate of 6.7 L h^{-1} at 3600 rpm [33]. The roller/crimper is 1.22 m wide and would cover approximately 0.67 ha h^{-1} operating at a speed of 6.44 km h^{-1} with 75% field efficiency [34]. Based on these parameters and gasoline physical properties (density) [35], the total gasoline consumption is 7.2 kg ha^{-1}, which is more than the total fuel/lubricant usage of 4.8 kg ha^{-1} reported by European researchers in Italy [36]. This difference is related to the 27 percent lower energy value for gasoline than from diesel fuel [35] along with the 75% field efficiency adjustment to account for overlap and turning around after each pass.

Using the procedures outlined in [34], both fixed and direct machinery costs for the tractor and roller/crimper were included in the economic calculations per pass with the roller/crimper. Total cost (fixed and direct) was US \$56.01 ha^{-1} for a single pass with the roller/crimper at this travel speed and would take approximately 1.5 h to complete a rolling one hectare of cover crop area. Rolling two and three times would take 3 and 4.5 h to complete at a cost of US \$112.02 and US \$168.02 ha^{-1}, respectively.

According to the USDA National Agriculture Statistics Service [37], the 5-year average retail price for fresh market bush beans was USD 1.36 kg^{-1}. Based on bush bean yield (Table 11), it would provide USD 27,569.92 ha^{-1}, USD 28,482.48 ha^{-1}, and USD 28,852.40 ha^{-1} income for 1, 2, and 3 rolling passes, respectively. An increase of USD 912.56 ha^{-1} and USD 1282.48 ha^{-1} would be given with two passes and three passes of rolling, respectively, compared to a single pass in income per ha^{-1}. However, since bush bean yield difference among rolling passes is not statistically different, benefits from rolling three times v/s once or twice are related to cover crop termination results. Cover crop termination data suggests that rolling three times could provide optimum conditions (above 90% termination rate and increased soil moisture) for planting the cash crop one week sooner than rolling one and two times. That additional week of planting opportunity could result in faster cash crop establishment and ultimately increased crop yield which could overcome the additional costs (e.g., fuel, labor, depreciation) to perform the multiple rolling/crimping operations. Our calculated additional cost of each pass does not include the opportunity cost of the producer's time. Some diversified mixed vegetable producers may have draws on their time from the other crops that they manage worth more than the additional profit we estimated for fresh market bush beans.

4. Conclusions

Cover crop termination rate was significantly higher for the rolling/crimping treatments compared to the non-rolled control. For all three years, the three times rolled treatment had significantly higher kill percentage compared to all other treatments at two weeks after rolling treatments were applied, with over 90% in 2020. This shows an advantage to rolling/crimping three times to allow successful cash crop that could potentially be performed one week earlier compared to the recommended 3-week interval, under certain conditions. The advantages of planting a week earlier are important in no-till to avoid increased weed and pest pressure as well as higher temperatures that occur later in the season. This could help get crops to market earlier. Rolling/crimping proved to be effective as yield was significantly higher compared to not rolled when averaged over all 3 growing seasons. Although statistical significance was not observed for yield and monetary benefits, rolling three times could provide better planting conditions and shorter harvest times by being able to sow earlier compared to the other rolling treatments. Difficulty planting into a standing cover crop exists which negatively effects seed to soil contact, but more importantly the higher soil moisture advantage on the rolled plots over the standing (untreated) cover crop plots was an important advantage of rolling/crimping. Soil moisture, when first planting, is the key to successful germination and establishing a good crop stand. Greater termination rates of the cover crop mixture resulted in better soil conditions for planting such as higher soil moisture and more brittle cover crop residue to be effectively cut and parted away from the planting path.

Author Contributions: T.S.K. and C.M.K. collaborated on experimental conceptualization, experimental investigation, contributing to resources, statistical analysis, and writing original draft preparation. All authors have read and agreed to the published version of the manuscript.

Funding: This research received no external funding and was funded by the USDA Agricultural Research Service.

Data Availability Statement: Not applicable.

Acknowledgments: Authors would like to acknowledge Trent Morton, (Agricultural Economist) for his help with the economic analysis of this study.

Conflicts of Interest: The authors declare no conflict of interest.

References

1. Ashford, D.L.; Reeves, D.W. Use of a mechanical roller crimper as an alternative kill method for cover crop. *Am. J. Altern. Agric.* **2003**, *18*, 37–45. [CrossRef]
2. Kornecki, T.S.; Price, A.J.; Raper, R.L. Performance of different roller designs in terminating cover crops. *Appl. Eng. Agric.* **2006**, *22*, 633–641. [CrossRef]
3. Kornecki, T.S.; Price, A.J.; Raper, R.L.; Arriaga, F.J. New roller crimper concepts for mechanical termination of cover crops. *Renew. Agric. Food Syst.* **2009**, *24*, 165–173. [CrossRef]
4. Kornecki, T.S.; Price, A.J. Effects of different roller/crimper designs and rolling speed on rye cover crop termination and seedcotton yield in a no-till system. *J. Cotton Sci.* **2011**, *14*, 212–220.
5. Kern, J.S.; Johnson, M.G. Conservation Tillage Impacts on National Soil and Atmospheric Carbon Levels. *Soil Sci. Soc. Am. J.* **1993**, *57*, 200–210. [CrossRef]
6. McGregor, K.C.; Mutchler, C.K. Soil loss from conservation tillage for sorghum. *Trans. ASAE* **1992**, *35*, 1841–1845. [CrossRef]
7. Reeves, D.W. Advances in Soil Science: Crops Residue Management. In *Cover Crops and Rotations*; Hatfield, J.L., Stewart, B.A., Eds.; Lewis Publishers: Boca Raton, FL, USA, 1994; pp. 125–172. [CrossRef]
8. Raper, R.L.; Reeves, D.W.; Burmester, C.H.; Schwab, E.B. Tillage depth, tillage timing, and cover crop effects on cotton yield, soil strength, and tillage energy requirements. *Appl. Eng. Agric.* **2000**, *16*, 379–385. [CrossRef]
9. Raper, R.L.; Reeves, D.W.; Schwab, E.B.; Burmester, C.H. Reducing soil compaction of Tennessee Valley soils in conservation tillage systems. *J. Cotton Sci.* **2000**, *4*, 84–90.
10. Ciaccia, C.; Canali, S.; Campanelli, G.; Testani, E.; Montemurro, F.; Leteo, F.; Delate, K. Effect of roller-crimper technology on weed management in organic zucchini production in a Mediterranean climate zone. *Renew. Agric. Food Syst.* **2016**, *31*, 111–121. [CrossRef]
11. Kornecki, T.S.; Balkcom, K.S. Organic Kale and Cereal Rye Grain Production Following a Sunn Hemp Cover Crop. *Agronomy* **2020**, *10*, 1913. [CrossRef]
12. Kornecki, T.S.; Price, A.J. Management of high-residue cover crops in a conservation tillage organic vegetable on-farm setting in Alabama. *Agronomy* **2019**, *9*, 640. [CrossRef]
13. Kornecki, T.S. Multistage Crop Roller. U.S. Patent 7,987,917 B1, 2 August 2011.
14. Kornecki, T.S.; Kichler, C.M. Modular Device for Cutting Cover Crop Residue. U.S. Patent US 10,959,363 B2, 30 March 2021.
15. Kornecki, T.S.; Arriaga, F.J.; Price, A.J. Evaluation of methods to assess termination rates of cover crops using visual and non-visible light active sensors. *Trans. ASABE* **2012**, *55*, 733–741. [CrossRef]
16. AWIS. Agricultural Weather Information Service, Inc. 2021. Available online: http://www.awis.com/mesonet/index.html (accessed on 4 November 2021).
17. Gomez, K.A.; Gomez, A.A. *Statistical Procedures for Agricultural Research*, 2nd ed.; John Wiley & Sons, Inc.: New York, NY, USA, 1984.
18. SAS. *Proprietary Software Release 9.2.*; SAS Institute, Inc.: Cary, NC, USA, 2013.
19. Kornecki, T.S.; Arriaga, F.J. Impact of different cover crops and types of transplanter mounted subsoiler shanks on tomato yield. *HortScience* **2011**, *46*, 715–720. [CrossRef]
20. Kornecki, T.S.; Price, A.J.; Balkcom, K.S. Cotton Population and Yield Following Different Cover Crops and Termination Practices in an Alabama No-Till System. *J. Cotton Sci.* **2015**, *19*, 375–386.
21. Kornecki, T.S. Effects of Different Rollers and Rye Termination Methods on Soil Moisture and Cotton Production in a No-Till System. *J. Cotton Sci.* **2020**, *24*, 197–210. [CrossRef]
22. Hodgskiss, C.L.; Young, B.G.; Armstrong, S.D.; Johnson, W.G. Evaluating cereal rye and crimson clover for weed sup-pression within buffer areas in dicamba-resistant soybean. *Weed Technol.* **2021**, *35*, 404–411. [CrossRef]
23. Vann, R.A.; Reberg-Horton, S.C.; Edmisten, K.L.; York, A.C. Implications of Cereal Rye/Crimson Clover Management for Conventional and Organic Cotton Producers. *Agron. J.* **2018**, *110*, 621–631. [CrossRef]
24. Reberg-Horton, S.C.; Grossman, J.; Kornecki, T.S.; Meijer, A.D.; Price, A.J.; Place, G.T.; Webster, T.M. Utilizing cover crop mulches to reduce tillage in organic systems in the southeast. *Renew. Agric. Food Syst.* **2012**, *27*, 41–48. [CrossRef]
25. Kornecki, T.S.; Arriaga, F.J.; Price, A.J.; Balkcom, K.S. Effects of recurrent rolling/crimping operations on cover crop termination, soil moisture, and soil strength for conservation organic systems. *Appl. Eng. Agric.* **2013**, *29*, 841–850. [CrossRef]

26. Kornecki, T.S. Rye termination by different rollers/crimpers developed for no-till small-scale farms. *Appl. Eng. Agric.* **2015**, *31*, 849–856. [CrossRef]
27. Kornecki, T.S. Influence of Recurrent Rolling/Crimping on Cover Crop Termination, Soil Strength and Yield in No-Till Cotton. *AgriEngineering* **2020**, *2*, 631–648. [CrossRef]
28. Kornecki, T.S.; Kichler, C.M. Effectiveness of Cover Crop Termination Methods on No-Till Cantaloupe. *Agriculture* **2022**, *12*, 66. [CrossRef]
29. Biuk, A.A. Physiological and Yield Responses of Snap Beans (*Phaseolus vulgaris*) to Water Availability. Ph.D. Thesis, Oregon State University, Corvallis, OR, USA, 8 December 1982.
30. Priyadarshini, V.M.; Madhanakumari, P. Effect of biostimulants on the yield of bush bean (*Lablab purpureus* var. *typicus*). *Ann. Plant Soil Res.* **2021**, *23*, 66–70. [CrossRef]
31. Hefner, M.; Canali, S.; Willekens, K.; Lootens, P.; Deltour, P.; Beeckman, A.; Kristensen, H.L. Termination method and time of agro-ecological service crops influence soil mineral nitrogen, cabbage yield and root growth across five locations in northern and western Europe. *Eur. J. Agron.* **2020**, *120*, 126144. [CrossRef]
32. Heathrow, F.L. American Automobile Association. Available online: https://gasprices.aaa.com/ (accessed on 19 August 2022).
33. Honda Motor Co., Ltd. Tokyo, Japan. Available online: https://www.hondapowerequipment.co.nz/engines/gx-big-series/gx690 (accessed on 19 August 2022).
34. Ag Decision Maker: Estimating Farm Machinery Costs. Iowa State University, Extension and Outreach, File A3-29, PM 710 Revised May 2015. Available online: https://www.extension.iastate.edu/agdm/crops/pdf/a3-29.pdf (accessed on 1 September 2022).
35. Viswanathan, B. Chapter 2—Petroleum. In *Energy Sources*, Fundamentals of Chemical Conversion Processes and Applications; Elsevier: Cambridge, MA, USA, 2017; pp. 29–57. [CrossRef]
36. Canali, S.; Campanelli, G.; Ciaccia, C.; Leteo, F.; Testani, E.; Montemurro, F. Conservation tillage strategy based on the roller crimper technology for weed control in Mediterranean vegetable organic cropping systems. *Eur. J. Agron.* **2013**, *50*, 11–18. [CrossRef]
37. USDA National Agricultural Statistics Service Quick Stats Ad-Hoc Query Tool. Available online: https://quickstats.nass.usda.gov/results/25EE941F-D345-3D20-BD57-34F8D5D73FB9 (accessed on 1 September 2022).

AgriEngineering

MDPI

Article

Increasing the Durability of Tools for Forest Road Maintenance

Monika Vargová [1], Łukasz Bołoz [2], Miroslava Ťavodová [1] and Richard Hnilica [1,*]

1 Department of Manufacturing Technology and Quality Management, Faculty of Technology, Technical University in Zvolen, T.G. Masaryka 24, 96001 Zvolen, Slovakia
2 Department of Machinery Engineering and Transport, Faculty of Mechanical Engineering and Robotics, AGH University of Science and Technology, A. Mickiewicza Av. 30, 30-059 Krakow, Poland
* Correspondence: hnilica@tuzvo.sk

Abstract: To ensure the care of forests, it is necessary to make them sufficiently accessible by forest roads. The basic working tool are hammers, or round shanks of various shapes, composed of a body and a tip. They are subject to a strong abrasive environment, which often leads to damage up to the complete destruction of the functional part of the tool. For these reasons, it is necessary to deal with the possibilities for increasing their lifetime. One of the possibilities of increasing the service life of these tools is hardfacing by welding. The article deals with the abrasive resistance of the original material of the tool and the hardfacing materials. Based on the chemical analysis of the base material of the tool, we found that the tool is made of manganese steel 38Mn6. This material was used as a standard and was compared with the hardfacing materials Abradur 58, E DUR 600, UTP DUR 600 and OK 84.58. Electron microscopy was used to evaluate the microstructure. Next, the Rockwell hardness measurement was performed on the samples. The original tool material 38Mn6 reached the lowest hardness value, namely, 21 HRC. The highest value was reached by the hardfacing material E DUR 600, namely, 59 HRC. Subsequently, a test of resistance to abrasive wear was performed according to GOST 23.208-79. Based on this test, we can conclude that the highest value of resistance to abrasive wear was achieved by Abradur 58. Even though the hardness of this coating was slightly lower than the hardfacing material E DUR 600, specifically 56 HRC, we can state that this hardfacing material (Abradur 58) achieved the best results among the investigated materials.

Keywords: forest road maintenance; tools durability; road milling machines; cutting tools; road cutters; road cutters; resistance to abrasive wear

check for updates

Citation: Vargová, M.; Bołoz, Ł.; Ťavodová, M.; Hnilica, R. Increasing the Durability of Tools for Forest Road Maintenance. *AgriEngineering* **2023**, *5*, 566–580. https://doi.org/10.3390/agriengineering5010036

Academic Editors: Marcello Biocca and Roberto Fanigliulo

Received: 13 January 2023
Revised: 24 February 2023
Accepted: 28 February 2023
Published: 7 March 2023

1. Introduction

To ensure the care of forests, it is necessary to make them sufficiently accessible by forest roads. Unfavorable terrain conditions and the relatively large weight of wood place high demands on the technical level of forest roads. Forest roads help to ensure the timely fulfillment of economic tasks related to logging, forest protection, cultivation, transport of harvested wood and temporary storage. Therefore, it is necessary to improve, reconstruct and maintain their high technical condition. Currently, road cutters-stone crushers are used for the construction and reconstruction of forest roads. The basic working tool are hammers, or round shanks of various shapes, composed of a body and a tip. The tools are exposed to difficult working conditions caused by the heterogeneous composition of the working environment, which mainly consists of hard rocks of various shapes and sizes. They are subject to a strong abrasive environment, which often leads to damage, up to the complete destruction of the functional part of the tool. There are frequent tool changes, which increases the working time and increase the cost of fuel, and the purchase of new tools. For these reasons, it is necessary to deal with the possibility of increasing their lifetime.

Currently, conical picks are the essential tools used in cutterheads of many working machines in various industries [1–3]. They are used in underground and opencast mining and often in construction, tunneling, road construction and maintenance. The service life of conical picks depends on their working conditions, where the most important is the abrasiveness of the mined rocks. In harsh conditions, these tools work for only a few hours, and changing them is time-consuming and requires the machine to be stopped. Hence results in a decrease in efficiency and higher costs of the process. Figure 1 shows an example of worn conical picks. Figure 1a,c have asymmetrical forms, and the rest have symmetrical forms. Typical cutterheads are usually equipped with about 40–60, but their number exceeds a hundred for longer drums. Therefore, their durability is crucial, and it has an economic dimension.

Conical picks are often the subject of research and development works in many universities and research centers worldwide. Research is often conducted to increase the durability of knives operating in abrasive conditions, i.e., to develop tools resistant to abrasive wear. There are many examples of such studies. In one of the articles, tools with a body protected with wear-resistant coatings and sintered carbide rings were tested [4]. In another, the mechanism of abrasive wear was studied, and tool wear prediction was proposed [5]. In the next one, the possibility of supporting the mining process was investigated [6]. Tests were also carried out for the cemented carbides themselves [7]. Complete tools [8] and entire cutting heads [9] are often tested. The following articles concerned the adaptation of modern tools and machines to difficult conditions [10,11] and the use of disc tools as an alternative to conical picks [12,13]. Testing the quality of tools was also discussed to facilitate the selection of the best offer in public tenders [14].

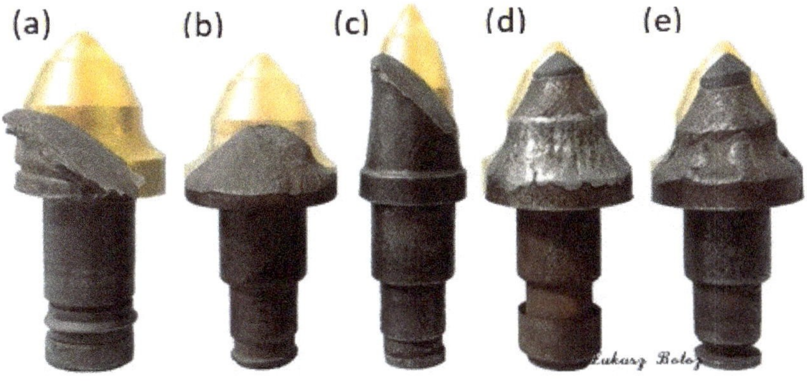

Figure 1. Worn conical picks [14]: (**a**,**c**) asymmetrical forms; (**b**,**d**,**e**) symmetrical forms.

A typical conical pick has a characteristic shape. It is made in the form of a solid revolution (Figure 2a). It consists of a working part (cutting part) 2 and a gripping part 3 (mounted in pick holder). The working part is reinforced with a sintered carbide insert 1. Conical picks are mounted in pick holders with special locks. The characteristic shape of the body and the mounting method allows free rotation of picks. The rotation of picks results in even wear of the working part and the insert. Thanks to the even wear, the picks shorten but retain their shape and can properly carry out the mining process. The body of the pick and the gripping part are made of steel characterized by high impact strength (usually $U > 25$ Jcm^{-2}) and very high tensile strength ($R_m > 1000$ MPa), as well as resistance to abrasive wear. The working part should have a hardness of at least 45 HRC, while the hardness of the gripping part should be in the range of 25 HRC–35 HRC [15]. Depending on working conditions, various types of steel are applied, such as 14NC11, 41Cr4, 40NiCr6, 36CRNIMo4 or 34CRMo4 (designations according to EN 10084). In addition, the knives are subjected to heat treatment to increase their hardness and, thus, the abrasion resistance of the surface layer. Additional wear-resistant layers are made of stellites or cemented carbides, usually based on cobalt, nickel and iron. These layers are made by hardfacing with electrodes. The hardness of the coatings may exceed 60 HRC [15].

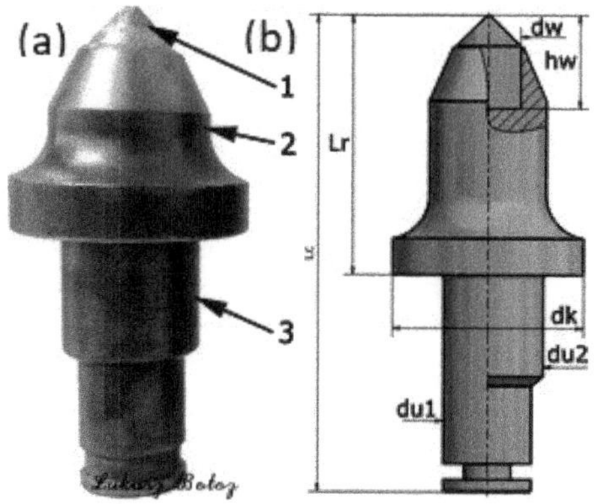

Figure 2. The most common conical picks: (**a**) scheme of the pick, (**b**) dimensions of the pick: 1—WC tip; 2—body; 3 — tool shank.

The inserts are made by sintering. Typically, cemented carbides for rock mining consist of wolfram carbide WC (89–95%), and the rest is cobalt C, a matrix. Wolfram carbide is hard and wear-resistant but brittle. Cobalt is the bonding phase and increases the toughness of the insert. The hardness of the cemented carbides exceeds 1050 HV30. Nowadays, sometimes users also require a specific grain size of WC. The inserts are soldered in sockets of working parts.

In addition to the material parameters discussed above, geometric and kinematic parameters also determine the correct mining course. Below is the range of parameters of typically tapered cutters used for mining rock and other materials. The markings are shown in Figure 2b.

Usually, these parameters amount to the following [15]:

- Length of pick: L_c = 120 mm–250 mm;
- Length of working part: L_r = 40 mm–100 mm;
- Gripping part diameter: d_u = ϕ20 mm–ϕ40 mm;
- Flange diameter: d_k = ϕ45 mm–ϕ70 mm;
- Mounting method: Seger, HERT, expanding or friction ring;
- Insert diameter: d_w = ϕ10 mm–ϕ25 mm;
- Insert height: h_w = 14 mm–40 mm;
- Tip angle: $2\beta_u$ = 80°–95° (more than 95° for ballistic shape);
- Yip shape: conical, multi-conical, ballistic, hat-shaped.

Many models of picks are available on the market, differing in material and geometrical parameters, shape and assembly method. This results in the existence of more than two hundred models of picks. Shape and size depend on the type of machine, which is related to the type of cutting material and environment (Figure 3a–d).

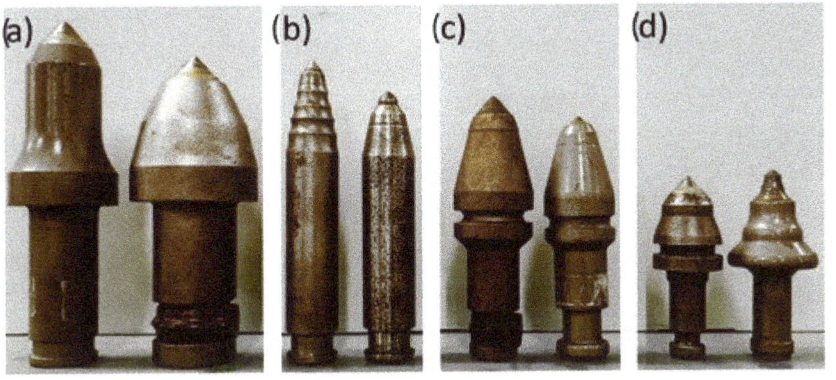

Figure 3. Often used types of conical picks: (**a**) for coals and rocks; (**b**) for salts; (**c**) for road, rocks, concrete; (**d**) for roads, concrete and asphalt.

Basic equipment for the creation, modification and reconstruction of forest roads includes adapters as additional equipment for UKT (universal wheeled tractor). Alternatively, they are designed as adapters for various special machines or mounted on hydraulic manipulators (Figure 4).

(**a**) (**b**) (**c**)

Figure 4. Machines for the creation, modification and reconstruction of forest roads: (**a**) for UKT; (**b**) for special machines; (**c**) for hydraulic manipulators.

Since the tools work in a highly abrasive environment, they are subject to a lot of abrasive wear. Abrasive wear occurs whenever a solid object is loaded with material particles of equal or greater hardness [16]. The abrasive wear mechanism is a complex process in the context of many factors. The intensity of these factors depends on the operating conditions of the environment in which these components and tools work. Furthermore, there are the operating parameters of the machines and the material properties of the contact surfaces [17]. There are several methods for increasing the resistance to abrasive wear. One such method is to apply additional materials to the exposed parts of the functional surfaces of the tools. Hardfacing is a commonly used method to improve the surface properties of agricultural tools, mining components and soil preparation equipment, among others [18]. In order to choose a suitable hardfacing material, it is necessary to know not only the basic material of the tool but also the environment in which the tool works.

2. Materials and Methods

The functional surfaces of the tool are the surfaces that are most involved in stone crushing and are exposed to strong abrasion and abrasive wear. In Figure 5, we see a new (unused) tool (Figure 5a) and a worn tool (Figure 5b) that has been used in continuous operation for approximately three months. We can see that there is a change in the shape and loss of material of the wolfram carbide (WC) tip and the area under the WC tip. From this, we can conclude that these surfaces are most involved in stone crushing. The tool had to be taken out of service after a high rate of wear.

Figure 5. Destruction of the working tool: (**a**) new tool; (**b**) worn tool.

We used the visualization method to detect the wear of the working tool. With this method, a 3D image (scan) of the new (Figure 6a) and the used (Figure 6b) working tool was performed. After subsequently overlaying them, we see that there was a change in the shape of the WC tip and a loss of material below the carbide tip (Figure 6c). This also confirmed the defined functional areas of the working tool.

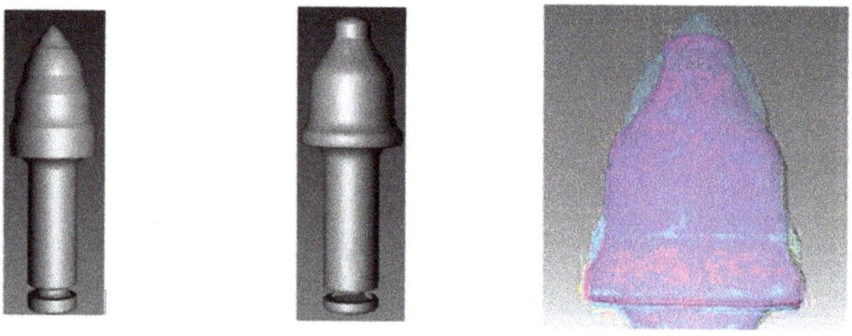

(**a**) (**b**) (**c**)

Figure 6. 3D image (scan) of the working tool: (**a**) 3D image (scan) of the new working tool; (**b**) 3D image (scan) of the working tool used; (**c**) overlay of the surfaces of the new and used working tool.

As it was not known what material the tools were made of, a chemical analysis was carried out. Based on the chemical analysis performed, we found that the tool is made of 38Mn6 manganese steel [19]. The chemical composition of this steel is in Table 1.

Table 1. The chemical composition of the base material.

Element	**C**	**Mn**	**Cr**	**Si**	**S**	**P**	**Ni**	**Mo**	**Fe**
(wt.%)	0.34–0.42	1.4–1.65	max. 0.4	0.15–0.45	max. 0.035	max. 0.035	max. 0.4	max 0.1	balance

The working tools of a road milling machine are subject to a lot of abrasive wear and need to be replaced frequently. Their frequent replacement causes technical and economic problems. For this reason, it is necessary to devise ways of increasing the lifetime of such tools. On the basis of practical experience, as well as the results of some authors [20,21], we have decided to hardfacing by welding the functional surfaces of the working tools of road milling machines.

To perform the experiment, we chose the following four types of electrodes:

- Hard deposit created by the ABRADUR 58 electrode [22];
- Hard deposit created by the E DUR 600 electrode [23];
- Hard deposit created by the UTP DUR 600 electrode [24];
- Hard deposit created by the WEARTHRODE 55 HD (OK 84.58) electrode [25].

The samples were made by a certified person for welding in the company ZOŠ Zvolen (Railway Repair and Engineering Works Zvolen). Hardfacing by welding was carried out manually by electric arc at welding position PA.

Sample 1 (Figure 7a) was made from the hard deposit formed with the ABRADUR 58 electrode. ABRADUR 58 is an electrode from SIJ ELECTRODE JESENICE, which creates a hard layer with extreme resistance to abrasion and a moderate impact. The hard deposit is made of chrome steel. Depending on the carbon and chromium content, the hard deposit has ferritic, austenitic-martensitic and semi-ferritic structures. It is mainly used for the hardsurfacing by the welding of crushers, parts of earth-moving machines and soft ore crushers. The typical hardness of the hard deposit is approximately 59 [20]. The chemical composition of the electrode can be seen in Table 2. For Sample 1 (Figure 7a), the electrodes were dried in a dryer at 300 °C before hardfacing by welding. Drying took 2 h. The preheating temperature of the base material was 150 °C. The ignition of the electric arc was simple, instantaneous hardfacing by welding occurred. An electrode with a diameter of Ø 2.5 mm was used for hardfacing by welding. The hard deposit was carried out on the base material of the tool. The hardfacing by welding conditions for Sample 1 are given in Table 3.

Figure 7. Samples: (**a**) Abradur 58; (**b**) E DUR 600; (**c**) UTP DUR 600; (**d**) OK 84.58.

Table 2. Basic data of the selected electrodes.

Electrode	Chemical Composition (wt. %)						Hardness (HRC)	Diameter Ø (mm)	Producer
	C	Cr	Si	Mn	P	Fe			
ABRADUR 58	3.2	32	-	-	-	balance	59	2.5	JESENICE
E DUR 600	0.5	8.5	-	-	-	balance	59	2.5	JESENICE
UTP DUR 600	0.5	9.0	2.3	0.4	-	balance	58	3.2	BÖHLER WELDING
OK 84.58	0.6	9.0	0.3	0.3	0.03	balance	57	3.2	ESAB

Table 3. Conditions of hardfacing by welding.

Conditions of Welding	Sample 1	Sample 2	Sample 3	Sample 4
Voltage (V)	22.5	21.5	21.5	21.5
Current (A)	100	140	140	140
Heat input	0.881	1.814	1.275	0.722

From the hardfacing by welding with electrode E DUR 600, Sample 2 (Figure 7b) was made. E DUR 600 is a hardfacing electrode from SIJ ELECTRODE JESENICE for hardfacing by welding parts that are exposed to abrasive wear associated with impacts. The hard deposit has a higher resistance to abrasion. The electrode is suitable for the surfaces of earth-moving machine parts and crushers. It is an electrode, forming the ledeburitic structure of the hard deposit, with a low content of carbon and chromium [23]. The chemical composition of the electrode can be seen in Table 2. The hardness of the hard deposit depends on the hardfacing by welding conditions and the chemical composition of the base material. The electrodes were dried at 400 °C for one hour before hardfacing by welding. The welded tool base material was preheated to 150 °C. The electric arc was unstable during hardfacing by welding. The electrode had a fragile shell. It was welded with an electrode with a diameter of Ø 2.5 mm. From Table 3, we can see that a lot of heat was introduced during the hardfacing by welding.

Sample 3 (Figure 7c) was made from the hard deposit formed with the UTP DUR 600 electrode. The UTP electrode DUR 600 is a universal electrode from BÖHLER WELDING. The hard deposit is abrasion, pressure and impact resistant, with a typical hardness of approximately 58 HRC. It is applied to armor parts of steels, manganese steels, castings and tool steels. Mainly used for earthmoving and construction machinery parts, hammer mills, crushing jaws and cones, hammer mills, etc. It has great hardfacing properties due to its quiet arc, easy slag removal, uniform current and good weld [24]. The basic data of the electrode are given in Table 2. The base material was preheated to 150 °C. The electrodes were dried in a dryer for two hours at 300 °C. Electrodes with a diameter of Ø 3.2 mm were used for hardfacing by welding. The hardfacing by welding conditions for Sample 3 are given in Table 3.

The OK WEARTRODE electrode (OK 84.58) from ESAB was made in Sample 4 (Figure 7d). The ESAB OK WEARTRODE electrode (OK 84.58) is an electrode for the hardfacing by welding of abrasive wear-resistant functional surfaces under simultaneous impact stresses with partial corrosion resistance. It is mainly used for parts of agricultural and forestry machinery, transport equipment, etc. The resulting hard deposit is formed by a martensitic structure. Full hardness is achieved already in the first layer of the hard deposit, regardless of the cooling rate [25]. The chemical composition of the electrode can be seen in Table 2. The sample base material was preheated to 150 °C. Electrodes with a diameter of Ø 3.2 mm were used for hardfacing by welding. The electrodes were dried in a dryer for 2 h at 200 °C. The hardfacing by welding conditions for Sample 3 are given in Table 3.

In Table 2 is the chemical composition and summarizes the basic data of all electrodes that were used for hardfacing by welding to the working tools of road milling machines.

Table 3 shows the hardfacing by welding parameters for all electrodes. The calculated heat input is also found here.

According to GOST 23.208-79-Testing the resistance of materials against wear by free abrasive particles (a group of standards ensuring the resistance of products against wear), all samples were prepared-base material 38Mn6 and deposits. The essence of the method described in the standard consists in comparing the loss of the tested material and the loss of the standard material under the same test conditions [26].

Due to the high hardness of the material, hydroabrasive cutting was used to cut the samples. The surface of the sample was milled and subsequently ground on a planar magnetic grinder to achieve dimensions of 30 mm × 30 mm × 10 mm with a roughness parameter Ra = 0.4 μm.

Vickers and Rockwell hardness measurement methods were chosen to evaluate the surface hardness of the samples. The hardness of the surface of the materials was measured in the laboratories of the Institute of Materials Research of SAV Košice. Vickers hardness was measured according to the procedure given in ISO 6507-1:2018 [27] on a Vickers 432SVD device. Load time t = 15 s and load force F = 98.07 N. Rockwell hardness was measured according to the procedure given in ISO 6508-1:2016 [28] on a UH250 device. The selected load force had a value of F = 1471 N.

The test of resistance to abrasive wear was performed according to GOST 23.208-79 [26]. The substance of the method consists of comparing the weight loss of the tested material and the standard material under the same test conditions. The grinding material used is electrocorundum with a grain size of 100–250 μm [29] with a relative humidity of φmax. = 0.15%. The hardness of electrocorundum corresponds to the 9th degree according to the Mohs scale. The standard [26] further states that when assessing wear resistance under specific wear conditions, it is possible to use an abrasive material corresponding to the material that acts during operation. However, the granularity condition must be maintained.

Before the test, each test body (standard, tested sample) is weighed and placed in the principle of the scheme of the test equipment for testing the abrasion resistance of the sample materials is shown in Figure 8. Each sample must be weighed on an accurate analytical balance before testing. Then the sample is placed in the holder of the testing equipment, the abrasive supply is started, and the rubber disc is pressed against the sample.

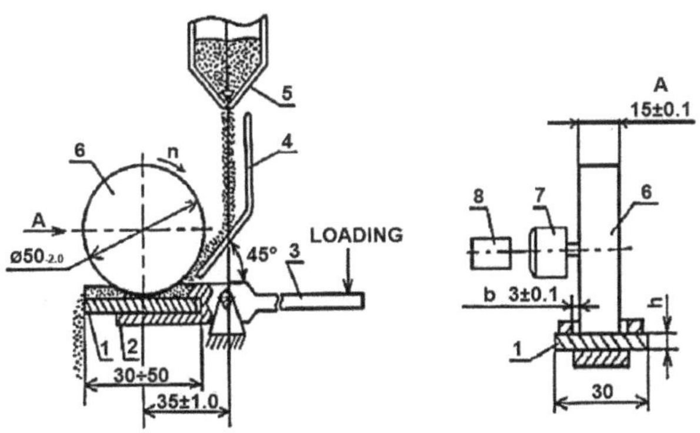

1. Sample
2. Sample holder
3. Working lever
4. Flow conditioner of abradant
5. Feed hopper
6. Rotate rubber cylinder
7. Driving gear
8. Device for checking the total cylinder speed during the test

n – rational speed
h – sample thickness

Figure 8. Scheme of the test equipment [26].

Three samples for each material were tested in the experiment. After each completed cycle, the sample was weighed three times on a Kern ABS analytical balance with a sensitivity of e = 0.1 mg. Arithmetic mean W_h was calculated for each measurement from the observed sample weight loss.

The test conditions were set as follows:

- Friction path length in one cycle R = 153.6 m;
- Rubber disc diameter D = 48.9 mm;
- Compressive force F = 15.48 N;
- Number of revolutions in one cycle n = 1000;
- Abrasive-silica sand OTTAWA with a grain size of 0.1 mm;
- Hardness of the abrasive material 54 HRC.

Samples were weighed after each cycle run. From the observed weight loss after each cycle, the arithmetic average for each sample was calculated.

The hardness coefficient K_T (-) is calculated from Formula (1) [26] as follows:

$$K_T = \frac{H}{H_a} \, (-) \tag{1}$$

where:

H—standard material hardness (HRC);

H_a—abrasive hardness (HRC).

The relative resistance to abrasive wear Ψ_h is calculated from the relation (2) [26] as'follows:

$$\Psi_h = \frac{W_{hE}}{W_{hPV}} \, (-) \tag{2}$$

where:

W_{hE}—mass loss of the standard sample (g);

W_{hPV}—mass loss of the tested sample (g).

3. Results and Discussion

In the metallographic analysis, we evaluated the microstructure of the base material of the working tool and the hardfacing materials. At the same time, we analyzed the interface of the base material of the working tool with the hardfacing material, their mutual mixing and the build-up zone during hardfacing by welding. A metallographic analysis was carried out at the Institute of Materials Research of the Slovak Academy of Sciences in Košice.

In Figure 9, we see the microstructure of the base material (BM). Cor etchant (120 mL CH_CCOOH, 20 mL HCl, 3 g picric acid, 144 mL CH_3OH) was used to induce the microstructure. The BM has a sorbitic microstructure-a mixture of ferrite and cementite. It is a ferritic-perlitic steel with a higher proportion of pearlite.

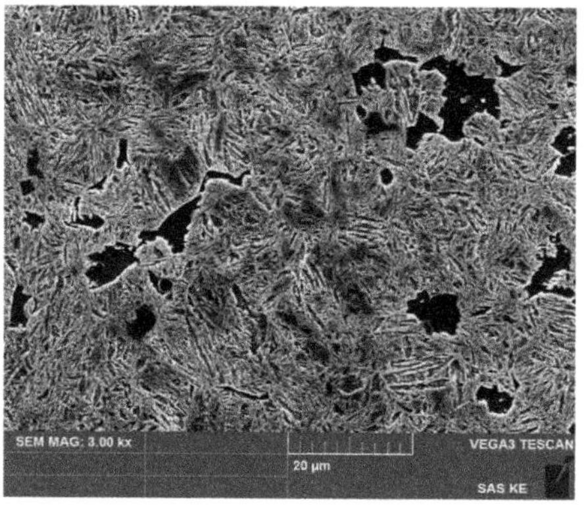

Figure 9. Microstructure BM–SEM.

In Figure 10a, we see the hardfacing material of Samples 1. Cor etchant was used to develop the microstructure. We also see the mixing of the base material with the facing material (Figure 10b). The hardfacing material is mixed without voids, cracks and other defects that adversely affect the quality of the hard deposit cohesion. The hardfacing material has a dendritic microstructure. We can see that the microstructure of the BM has changed in the heat-affected zone (HAZ). We observe the presence of tempered martensite.

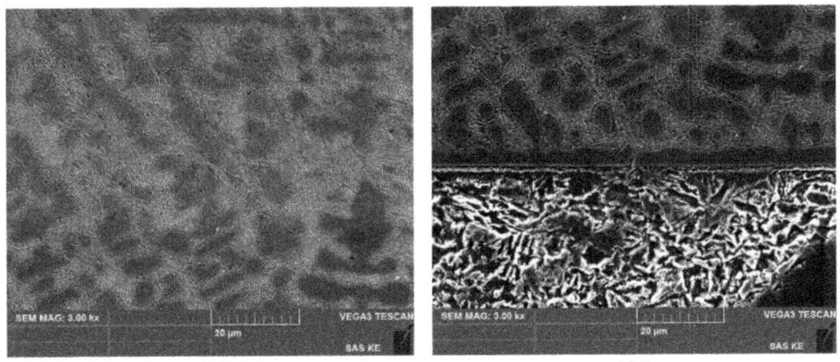

(**a**) (**b**)

Figure 10. Microstructure of Sample 1: (**a**) hardfacing material; (**b**) interface hardfacing material-BM.

In Figure 11a, we see the microstructure of the hardfacing material of Sample 2. The sample was etched with a Cor etcher. The hardfacing material has a ledeburitic microstructure. It is identical to the microstructure stated by the manufacturer. In Figure 11b, we can see the interface hardfacing material-BM. The connection of the BM and the hardfacing material is without defects and without breaking the integrity of the hardfacing material and the BM.

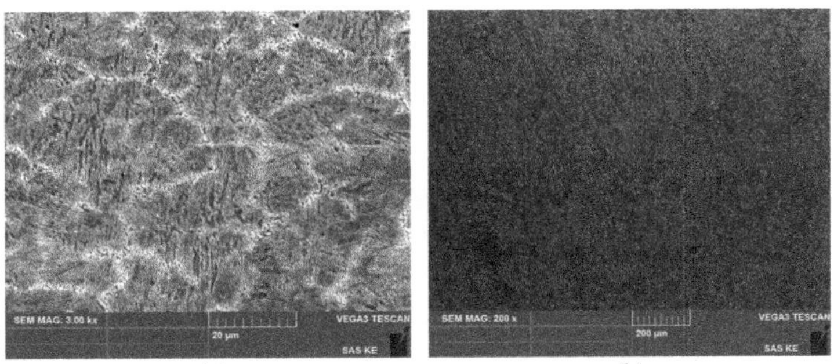

(**a**) (**b**)

Figure 11. Microstructure of Sample 2: (**a**) hardfacing material; (**b**) interface hardfacing material-BM.

In Figure 12a, we see the microstructure of the hardfacing material of Sample 3. The sample was etched with 2% Nital (solution of HNO_3 in ethyl alcohol). We observe the acicular needle-like structure of BM in the HAZ. In the HAZ, the presence of diffusion between the BM and the hardfacing material can be seen. We can conclude that there was a good mixing of the hardfacing material with BM (Figure 12b).

In Figure 13a, we see the microstructure of the hardfacing material of Sample 4. The sample was etched with a Cor etcher. We observe the polyhedral and acicular microstructure of BM in the HAZ (Figure 13b). We can see that the mixing of the hardfacing material with the BM of the tool is free of cracks, voids and other defects. There was a diffusion of chemical elements between the BM and the hardfacing material.

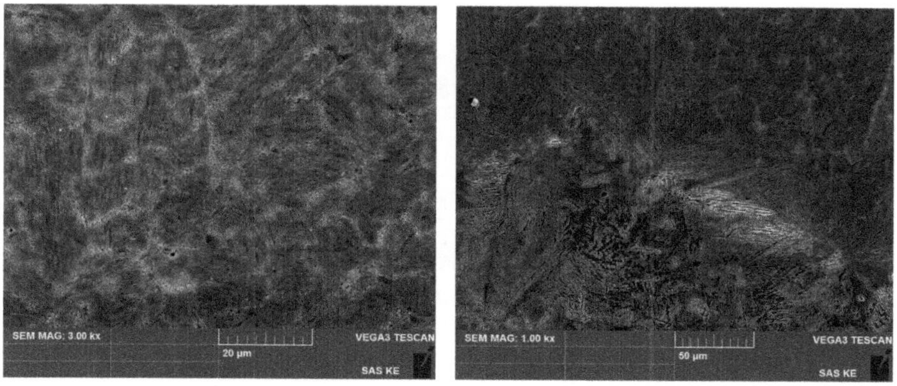

(a) (b)

Figure 12. Microstructure of Sample 3: (**a**) hardfacing material; (**b**) interface hardfacing material-BM.

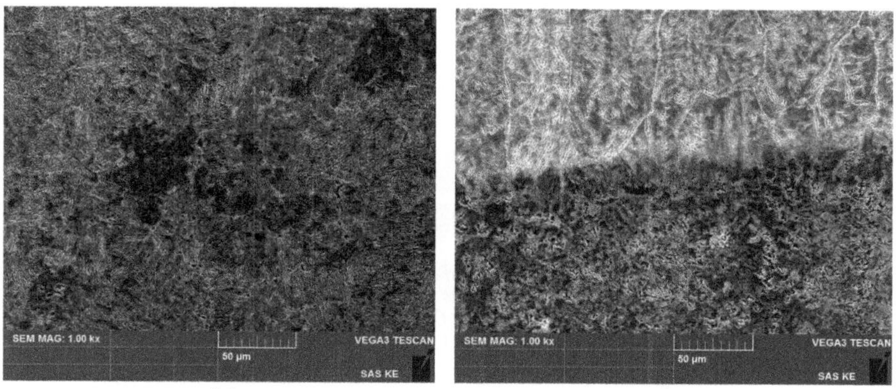

(a) (b)

Figure 13. Microstructure of Sample 4: (**a**) hardfacing material; (**b**) interface hardfacing material-BM.

The Vickers-HV0.5 measurement method and the Rockwell measurement method were used to measure the hardness of the base material samples and hardfacing material.

The indentations during measurement were guided from the core of the base material to the surface of the hardfacing material. The average hardness of BM, according to Vickers, was 285 HV0.5. The course of the measured hardnesses on Sample 1 is shown in Figure 14a. We can see that the thickness of the hardfacing material layer was about 3 mm. The hardness of the hardfacing material is significantly higher than the hardness of the base material of the working tool. The highest hardness value of the hardener was 652 HV0.5. The highest hardness value in the HAZ was 353 HV0.5.

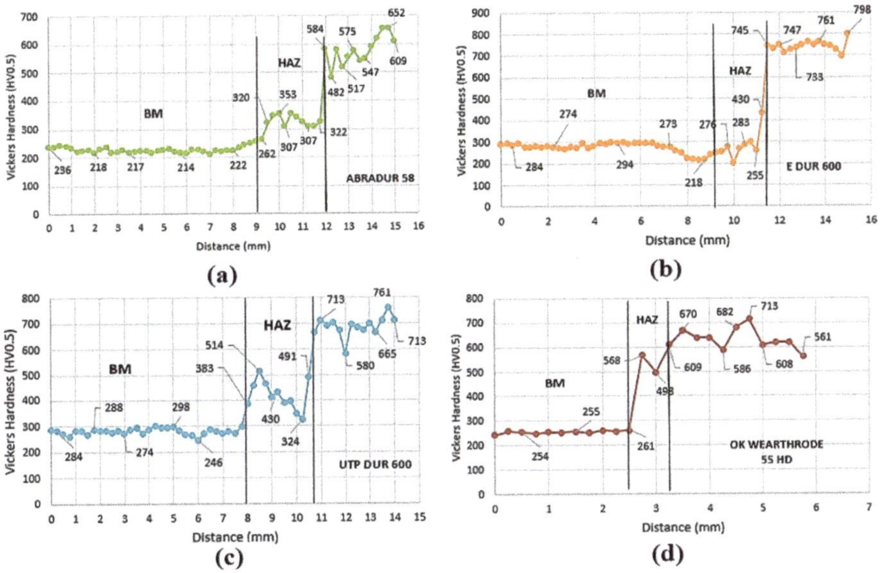

Figure 14. Graphic representation of the course of hardness according to Vickers: (**a**) Sample 1; (**b**) Sample 2; (**c**) Sample 3; (**d**) Sample 4.

The course of the measured hardnesses of Sample 2 can be seen in Figure 14b. We can see that the hardfacing material also has significantly higher hardness values than the base material of the tool. The hardness of Sample 2 is higher than the hardness of Sample 1. The hardfacing material layer of Sample 2 is about 4 mm thick. The highest hardness value in the HAZ was 430 HV0.5.

The course of the measured hardnesses in Sample 3 can be seen in Figure 14c. The highest hardness of Sample 3 hardfacing material had a value of 761 HV0.5. The hardfacing material layer is approximately 3 mm thick. We can see that in the HAZ, the hardness values fluctuate considerably, which could have been caused by uneven mixing of the hardfacing material with the tool base material at the point of measurement or by the structure of the material in the HAZ. The highest hardness value in the HAZ was 491 HV0.5.

The greatest hardness of the hardfacing material on Sample 4 has a value of 713 HV0.5, and in the HAZ, it has a value of 568 HV0.5. The hardfacing material layer has a thickness of approximately 2.5 mm. The course of the measured hardnesses on Sample 4 can be seen in Figure 14d.

The average measured values of samples and BM microhardness are shown in Table 4. We can see that the measured hardness values of hardfacing materials are significantly higher than the hardness of BM.

Table 4. Average values of microhardness of hardfacing materials and BM.

	BM	Sample 1	Sample 2	Sample 3	Sample 4
Average HV0.5	285	632	741	688	577

The average Rockwell hardness values of BM and hardfacing materials are shown in Table 5. We can conclude that all hardnesses of hardfacing materials are significantly higher than the hardness of the base material.

Table 5. Measured hardness values by the Rockwell method.

	BM	Sample 1	Sample 2	Sample 3	Sample 4
Average	21	56	59	58	52

Resistance to abrasive wear was tested according to GOST 23.208-79. For each tested material, three samples were tested. Before the test, the sample is weighed, placed in the device and the feed of the abrasive is started. After each completed cycle, the sample is removed and reweighed to determine the weight loss. The arithmetic average W_h is calculated from the detected weight losses.

First, the K_T hardness coefficient is calculated according to relation (1). This value will provide us with the first information about a possible better or worse abrasion resistance. Subsequently, relative resistance to abrasive wear Ψ_h is calculated according to relation (2). BM has a value of 1 because it is a standard sample. All obtained and calculated data from the test of resistance to abrasive wear are presented in Table 6.

Table 6. Values from the test of resistance to abrasive wear.

	K_T (—)	W_h (g)	Ψ_h (—)
BM	0.39	0.0996	1
Sample 1	1.04	0.0105	9.49
Sample 2	1.09	0.0247	4.03
Sample 3	1.07	0.0325	3.06
Sample 4	0.96	0.0355	2.80

In Figure 15a, we can see a graphic representation of the K_T hardness coefficients for individual materials. If this value is higher than 1, it is assumed that the tested material will be able to better resist abrasive particles. We can state that the highest value of the hardness coefficient was achieved by Sample 2. At the same time, the lowest value of the K_T hardness coefficient was achieved by BM, namely, 0.39. In Figure 15b, we can see a graphical comparison of values of relative resistance to abrasive wear Ψ_h. It is clear from the graph that Sample 1 achieved the best relative resistance to abrasive wear. It is almost 9.5 times more compared to the BM of the tool. Sample 4 achieved the smallest value of relative resistance to abrasive wear, namely, 2.8. However, it is 2.8 times more compared to the BM of the tool. We can conclude that all the tested hardfacing materials achieved several times better resistance to abrasive wear compared to the BM of the tool.

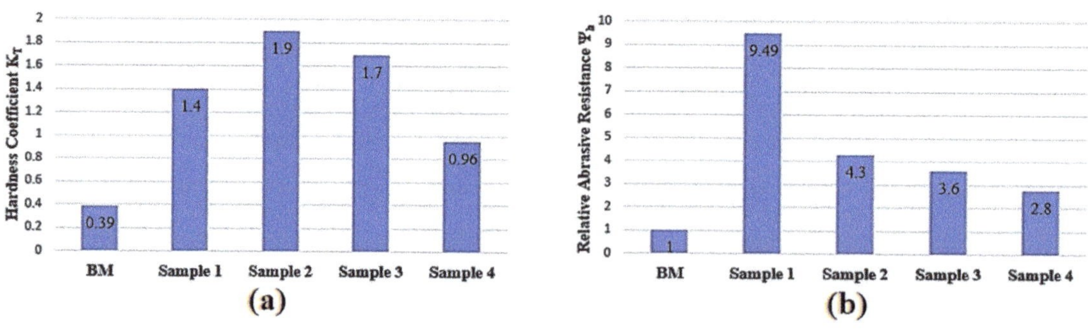

Figure 15. Graphic representation of the values from the abrasion resistance test: (**a**) hardness coefficient K_T; (**b**) relative abrasive resistance Ψ_h.

Some authors report that hardness is strongly correlated with resistance to abrasive wear [30–32]. However, based on the results of the hardness measurement and the test of resistance to abrasive wear, we can conclude that the hardfacing material with the highest hardness value (59HRC) achieved a lower value of relative abrasion resistance compared to the hardfacing material with lower hardness (56HRC). The hardfacing material with lower hardness achieved a 2.36 times higher value of relative resistance to abrasive wear. In the results of their research, the authors [18] also state that hardness is correlated with resistance to abrasive wear. However, from the partial results, it can be concluded that the W-rich alloy reached a lower value of resistance to abrasive wear compared to the Cr-rich hardfacing material, even though its hardness was higher than that of the Cr-rich hardfacing material.

4. Conclusions

Based on the chemical analysis of the base material of the tool, we found that the tool is made of manganese steel 38Mn6. Based on the results, it is possible to state the following:

- Based on the evaluation of the microstructure, we can conclude that there was a good mixing of BM with the hardfacing material for all four samples. Therefore, we can conclude that the selected hardfacing materials are suitable for hardfacing by welding on BM 38Mn6;
- Based on the results from the Rockwell hardness measurement and the test of resistance to abrasive wear, we can conclude that hardness, in this case, did not correlate with resistance to abrasive wear. The highest value of resistance to abrasive wear was achieved by the hardfacing material Abradur 58, despite the fact that its hardness was not the highest among the investigated hardfacing materials. The microstructure, which was dendritic with the presence of tempered martensite, could have had an effect on the resistance to abrasive wear.

Hardfacing by welding with an Abradur 58 electrode appears to be the most suitable solution for increasing the lifetime of road milling tools. Therefore, we recommend hardfacing by welding on the new tool this additional hardfacing material, which will strengthen its exposed surfaces. A tool modified in this way could better withstand the effects of an abrasive environment.

Author Contributions: M.V. conceptualization and research, Ł.B. theory and state of art and writing, M.Ť. methodology and research data analysis, R.H. results, review and editing. All authors have read and agreed to the published version of the manuscript.

Funding: This work was supported by "Agentúra na podporu výskumu a vývoja MŠVVaŠ SR" under the contract number APVV-16-0194 and "Operational Programme Integrated Infrastructure (OPII) funded by the ERDF" under the contract number ITMS 313011T720.

Data Availability Statement: Not applicable.

Conflicts of Interest: The authors declare no conflict of interest.

References

1. Dewangan, S.; Chattopadhyaya, S.; Hloch, S. Wear assessment of conical pick used in coal cutting operation. *Rock Mech. Rock Eng.* **2015**, *48*, 2129–2139. [CrossRef]
2. Wang, X.; Su, O.; Wang, Q.F.; Liang, Y.P. Effect of cutting depth and line spacing on the cuttability behavior of sandstones by conical picks. *Arab. J. Geosci.* **2017**, *10*, 525. [CrossRef]
3. Krauze, K.; Mucha, K.; Wydro, T. Evaluation of the quality of conical picks and the possibility of predicting the costs of their use. *Multidiscip. Asp. Prod. Eng.* **2020**, *3*, 491–504. [CrossRef]
4. Chang, S.-H.; Lee, C.; Kang, T.-H.; Ha, T.; Choi, S.-W. Effect of Hardfacing on Wear Reduction of Pick Cutters under Mixed Rock Conditions. *Geomech. Eng.* **2017**, *13*, 141–159.
5. Gajewski, J.; Jedliński, Ł.; Jonak, J. Classification of Wear Level of Mining Tools with the Use of Fuzzy Neural Network. *Tunn. Undergr. Space Technol.* **2013**, *35*, 30–36. [CrossRef]
6. Kotwica, K. The Influence of Water Assistance on the Character and Degree of Wear of Cutting Tools Applied in Roadheaders. *Arch. Min. Sci.* **2011**, *56*, 353–374.

7. Nahak, S.; Dewangan, S.; Chattopadhyaya, S. Discussion on Wear Phenomenain Cemented Carbide. *Procedia Earth Planet. Sci.* **2015**, *11*, 284–293. [CrossRef]
8. Liu, S.; Ji, H.; Liu, X.; Jiang, H. Experimental Research on Wear of Conical Pick Interacting with Coal-Rock. *Eng. Fail. Anal.* **2017**, *74*, 172–187. [CrossRef]
9. Krauze, K.; Bołoz, Ł.; Wydro, T. Parametric Factors for the Tangential-Rotary Picks Quality Assessment. *Arch. Min. Sci.* **2015**, *60*, 265–281. [CrossRef]
10. Bołoz, Ł.; Krauze, K. Ability to Mill Rocks in Open-Pit Mining. In Proceedings of the 18th International Multidisciplinary Scientific GeoConference SGEM2018, Albena, Bulgaria, 2–8 July 2018.
11. Bołoz, Ł.; Krauze, K. Disc Unit Dedicated to Mine Abrasive Rocks and in Particular Copper Ores. In Proceedings of the 18th International Multidisciplinary Scientific GeoConference SGEM2018, Albena, Bulgaria, 2–8 July 2018.
12. Gospodarczyk, P.; Kotwica, K.; Stopka, G. A New Generation Mining Head with Disc Tool of Complex Trajectory. *Arch. Min. Sci.* **2013**, *58*, 985–1006. [CrossRef]
13. Bołoz, Ł. Directions for Increasing Conical Picks' Durability. *New Trends Prod. Eng.* **2019**, *2*, 277–286. [CrossRef]
14. Bołoz, Ł. Results of a Study on the Quality of Conical Picks for Public Procurement Purposes. *New Trends Prod. Eng.* **2018**, *1*, 687–693. [CrossRef]
15. Bołoz, Ł.; Kalukiewicz, A.; Galecki, G.; Romanyshyn, L.; Romanyshyn, T.; Giménez, R.B. Conical Pick Production Process. *New Trends Prod. Eng.* **2020**, *3*, 231–240. [CrossRef]
16. Kovaříková, I.; Szewczyková, B.; Blaškoviš, P.; Hodúlová, E.; Lechovič, E. Study and characteristic of abrasive wear mechanisms. *Mater. Sci. Technol.* **2009**, *1*, 1–8.
17. Suchánek, J.; Kuklík, V.; Zdravecká, E. Influence of microstructure on erosion resistance of steels. *Wear* **2009**, *267*, 2092–2099. [CrossRef]
18. Buchely, M.F.; Gutierrez, J.C.; Leon, L.M.; Toro, A. The effect of microstructure on abrasive wear of hardfacing alloys. *Wear* **2005**, *259*, 52–61. [CrossRef]
19. Steel 38Mn6. Available online: http://www.Manusteelcn.Com/2013/09/38mn6.Html (accessed on 5 December 2022).
20. Müller, M.; Novák, P.; Chotěborský, R.; Hrabě, P. Reduction of Ploughshare Wear by Means of Carbide Overlay. *Manuf. Technol.* **2018**, *18*, 72–78. [CrossRef]
21. Vargova, M.; Tavodova, M.; Monkova, K.; Dzupon, M. Research of Resistance of Selected Materials to Abrasive Wear to Increase the Ploughshare Lifetime. *Metals* **2022**, *12*, 940. [CrossRef]
22. ABRADUR 58. Available online: https://hbt-weld.cz/app/uploads/2019/02/Odoln%C3%A9-proti-opot%C5%99eben%C3%AD-odd%C3%ADl-L.pdf (accessed on 5 December 2022).
23. E DUR 600. Available online: https://vecowelding.com/wp-content/uploads/2016/08/elektroda-jesenice-catalog-2009.pdf (accessed on 5 December 2022).
24. UTP DUR 600. Available online: http://resurszao.com/wp-content/uploads/2014/04/UTP-DUR-600.pdf (accessed on 5 December 2022).
25. OK 84.58. Available online: https://www.rapidwelding.com/files/8458504020.pdf (accessed on 5 December 2022).
26. Russian Standard: ГОСТ 23.208-79 Обеспечение износостойкости изделий. Метод испытания материалов на износостойкость при трении о нежестко закрепленные абразивные частицы. [In English: GOST 23.208-79 Ensuring of Wear Resistance of Products. Wear Resistance Testing of Materials by Friction Against Loosely Fixed Abrasive Particles]. Available online: https://www.internet-law.ru/gosts/gost/4066 (accessed on 8 December 2022).
27. *ISO 6507-1:2018*; Metallic Materials–Vickers Hardness Test–Part 1: Test Method. NSAI: Dublin, Ireland, 2018.
28. *ISO 6508-1:2016*; Metallic Materials–Rockwell Hardness Test–Part 1: Test Method. NSAI: Dublin, Ireland, 2018.
29. GOST 3647-80 Abrasives. Grain Sizing. Graininess and Fractions. Test Methods. Available online: Https://Docs.Cntd.Ru/Document/1200016841 (accessed on 8 December 2022).
30. Kováč, I.; Mikuš, R.; Žarnovský, J.; Drlička, R.; Harničárová, M.; Valíček, J.; Kadnár, M. Increasing the Wear Resistance of Surface Layers of Selected Steels by TIG Electric Arc Surface Remelting Process Using a Powder Based on CaCN2. *Int. J. Adv. Manuf. Technol.* **2022**, *123*, 1985–1997. [CrossRef]
31. Slota, J.; Kubit, A.; Gajdoš, I.; Trzepieciński, T.; Kaščák, Ľ. A Comparative Study of Hardfacing Deposits Using a Modified Tribological Testing Strategy. *Lubricants* **2022**, *10*, 187. [CrossRef]
32. Balla, J.; Mikuš, R.; Cviková, H. *Náuka o Materiáloch: (Návody Na Cvičenia)*; Slovenská Poľnohospodárska Univerzita: Nitra, Slovakia, 2003; ISBN 80-8069-217-3.

AgriEngineering

Article

Workability Assessment of Different Stony Soils by Soil–Planter Interface Noise and Acceleration Measurement

Pietro Toscano [1], Maurizio Cutini [1,*], Alex Filisetti [1], Elia Premoli [1], Maurizio Porcu [2], Nicola Catalano [2], Carlo Bisaglia [1] and Massimo Brambilla [1]

1 CREA Consiglio per la Ricerca in Agricoltura e L'analisi Dell'economia Agraria (Research Centre for Engineering and Agro-Food Processing), Via Milano 43, 24047 Treviglio, Italy
2 Codevintec Italiana srl—Tecnologie per le Scienze della Terra e del Mare, Via G. Labus 13, 20147 Milano, Italy
* Correspondence: maurizio.cutini@crea.gov.it

Abstract: Sowing is critical for successful crop establishment and productivity, particularly in precision agriculture management strategies. However, topsoil characteristics directly affect agribusiness maximization (i.e., crop-yield increase, machinery efficiency, operating-cost reduction) even in the most advanced farming management techniques. The excessive presence of coarse fractions or stones in arable soil layers prevents modern machinery from reaching optimal efficiency. This work focuses on sowing to verify whether the vibration and noise arising during this operation significantly change with varying soil conditions according to the stoniness degree of disturbance on soil workability. To make this assessment, an experimental sowing activity was carried out on four soil plots with two different disturbance degrees. The results confirmed that the noise and acceleration of the sowing machine significantly correlated with the soil disturbance degree and related workability profile.

Keywords: sustainable agriculture; precision farming; stoniness degree of disturbance; stony-soils workability; seeder machine; seeding speed

Citation: Toscano, P.; Cutini, M.; Filisetti, A.; Premoli, E.; Porcu, M.; Catalano, N.; Bisaglia, C.; Brambilla, M. Workability Assessment of Different Stony Soils by Soil–Planter Interface Noise and Acceleration Measurement. *AgriEngineering* **2022**, *4*, 1139–1152. https://doi.org/10.3390/agriengineering4040070

Academic Editor: Pablo Martín-Ramos

Received: 19 September 2022
Accepted: 13 November 2022
Published: 16 November 2022

1. Introduction

In modern agricultural management strategies, especially in precision-agriculture (PA) systems, sowing is one of the key factors influencing crop establishment and yield. Usually, before sowing, various tillage methods and types of machinery are used to prepare the soil, according to soil characteristics, farming techniques and crop needs, to destroy weeds and pests and improve the structural conditions of the arable layer, so as to maximize the efficiency of seed planters, and enhance seed germination and the growth of seedling roots. In the most advanced farming management typologies, maximizing arable farming outputs—in terms of crop yield, machinery efficiencies and reduction in operating costs—directly depends on topsoil quality. When defining and assessing the textural characteristics of arable layers, soil fractions over 2 mm in diameter (gravel and stones) are usually excluded because they are considered inert fractions.

However, an excessive presence of coarse fractions or stones in the soil's arable layers hinders or is incompatible with the operational requirements of modern cultivation techniques and machinery (i.e., minimal tillage, precision sowing), which require fields without obstructions and topsoil of fine earth to work at their best [1–3].

This work focuses on sowing to verify if the vibration and noise arising during such operation significantly change at varying soil conditions according to the stoniness disturbance degree on soil workability. The importance of vibration on the seed metering performance has been also highlighted for the design of a seed metering device and entire seeder, as well as the determination of operation speed [4,5].

1.1. Sowing Efficiency Issues: Precision Seeders

In the panorama of digital and precision farming machinery, seeders are a key tool for optimizing crop results. According to different sowing requirements, various types of precision seeders have been developed to meet the needs resulting from the geographical environment, farm scale, mechanization level and crop type of different countries, for both tilled and untilled soils [6,7].

Across their different types and structural and operational characteristics, such machines maximize sowing efficiency in terms of costs and time savings. Moreover, precise seed positioning along the row and at the proper sowing depth avoids seed wastage and fosters optimal crop growth.

Precision seeders are generally semi-portable machines consisting of a variable number of independent planter units fixed to the frame. Each element, therefore, constitutes a complete sowing unit, consisting of a hopper, distribution and damping devices, seed covers and a compaction device. In some cases, they can be combined with equipment for complete and minimal tillage or equipped with simple implements suitable for opening the furrow for depositing the seed in the firm soil (sod seeding).

Under optimal soil conditions, mechanical seeders operate at speeds between 5 and 7 km h^{-1} with power requirements varying between 1.5 and 2.0 kW per row. Pneumatic seeders can operate at speeds of up to 10–12 km h^{-1}, ensuring good uniformity of distribution, albeit with higher power requirements, typically 4–5 kW per row [8]. The planters are suitable for precision planting under a working speed of 6.9 km h^{-1} [9]. The most recent ISOBUS [10] models, with electrically driven seeding units, tractor auto-guidance system and global navigation satellite system (GNSS) positioning, can operate at speeds of up to 15 km h^{-1} with high levels of precision [11]. Various authors have evaluated the operational performance of different seeders in different areas and for different crops. However, neither bibliographic references nor technical documentation report correlations between machinery performances and soil condition, assuming that the soil is optimal for the performances of the described machines [12–15]. On the contrary, the type and structural conditions of the soil on which sowing occurs heavily affect the operational efficiency of seeders in terms of waving, humidity, and seedbed preparation methods: the presence of skeleton in the arable soil layer is widespread in a great part of the world's arable lands [16,17].

One of the main factors characterizing the efficiency of a seeder in terms of work rate, with the related costs for the tractor, operator, and fuel consumption, is the maximum allowable forward speed. Apart from technical considerations, such speed can depend on soil composition. For example, Figure 1a reports operation on soil with high skeleton which allowed a maximum speed of 5.6 km h^{-1}. Figure 1b reports the performance of the same tractor and seeder during an experimental sowing experience at various forward speeds on soil without skeleton, correctly sowing at 11.5 km h^{-1}. In this last case, the limit was the maximum power of the tractor used.

Research on sowing quality associated with high speeds [18,19] report operational limits due to excessive lateral soil throw, reduction in furrow backfill and interactions between adjacent furrows. In addition, they showed potential for new opener technology to increase operating speeds, increasing the timeliness of sowing, grain-yield potential and lowering total seeding time per season [20,21], encouraging the approach to technical improvements in the implement specific to particular soil conditions.

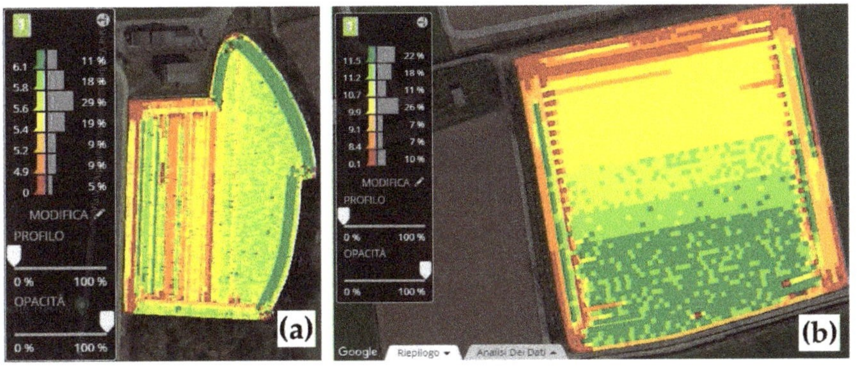

Figure 1. The forward speed of two different sowing operations on soil with different skeleton levels. (**a**) reports operation on soil with high skeleton which allowed a maximum speed of 5.6 km h^{-1}. (**b**) reports the same tractor and seeder on soil without skeleton correctly sowing at 11.5 km h^{-1}.

1.2. The Stony-Soils Disturbance Degree

The type and amount of coarse fractions define the workability class of soils as a function of the degree of disturbance (DD) caused to cultivation machinery (Figure 2) [22]. Soil stoniness can severely impair the efficiency of this machinery, in particular of seeders, in terms of both sowing effectiveness and structural integrity, even for those equipped with the most advanced vibration-damping systems and downforce control of sowing elements [23].

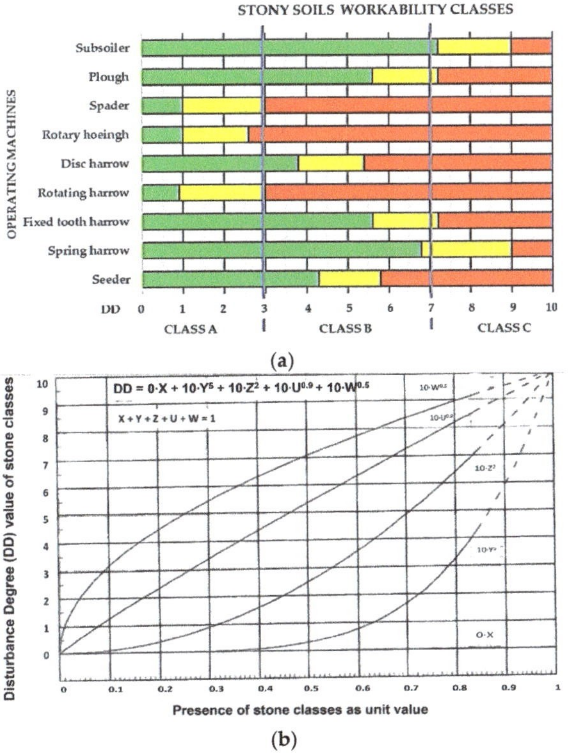

Figure 2. Stony-soil classes and workability limits of some operating machines (**a**), according to the stone classes disturbance degree (**b**) [22].

Understanding the effect of soil conditions on sowing performance in terms of precision and work rate requires the possibility of monitoring the parameters correlated with the operation.

The monitoring of some parameters is already adopted in rotating machines for the diagnostic analysis of the correct functioning. For example, vibrations and noise could be specific signals that indicate abnormal functioning [24]. Focused engineered sensors measuring vibrations have been developed for measurements on a hay-handling machine and have confirmed the capability of the online monitoring system to diagnose malfunctions. The proposed measuring system could also be used to improve the durability of agricultural machinery [25].

A purpose-designed array of instruments has been developed to install accelerometers and microphones for evaluating grain losses in key points of combined harvesters [26]. With this aim, this study focuses on the possibility of evaluating whether the level of vibration/acceleration or the noise of seeder units upon their impact with soil could be a signal correlated to the seeding quality or mechanical protection of the machine when operating on stony soils. Reducing vibration at a seeder distributor element to optimize sowing quality (uniform distribution) has also been developed as a patent [27].

This work aims to verify whether vibrations and noise during sowing can be measured and acquired and can identify significant changes in soil conditions according to the stoniness degree of disturbance on soil workability.

The results could also be discussed alongside the recent development of the ISOBUS function tractor implement management system (TIM). TIM is an ISOBUS-based agricultural technology system which enables the implement to control tractor functions such as, for example, forward working speed.

2. Materials and Methods

2.1. Soil Geoelectric Analysis/Test Plots

At the Council for Agricultural Research and Economics (CREA) experimental farm in Treviglio, Bergamo, Italy (45°31′14″ N; 9°35′27″ E; +128 m asl), in a soil classified as Calcic Skeletic Mollic Umbrisol, with neutral-sub alkaline pH, according to the International Union of Soil Science (IUSS) classification [28], a geoelectric analysis followed by the application of a classification algorithm on the 0–1 m deep soil's resistivity values resulted in the zoning of three homogeneous macro areas (Figure 3) [29].

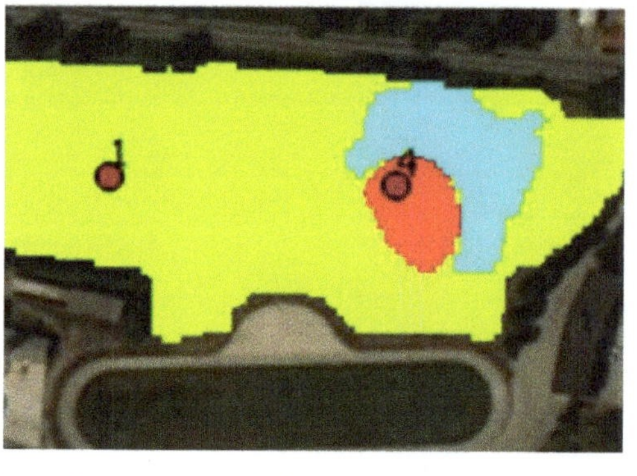

Figure 3. The shape of the homogeneous geoelectric soil zones.

Profile No. 4 is representative of soil class "A", consisting of soils with a depth of about 60 cm, with a medium sandy texture, characterized by a surface layer of about 20 cm of dark-brown color, with skeleton (25–35%) consisting of pebbles and gravels, of moderate subangular polyhedric, fine and medium structure and with fine pores. On the surface, there is a relatively abundant presence of pebbles. From a depth between 20 cm to about

60 cm, there is a brown-color layer, with abundant medium and large skeleton (50–60%) with weak fine subangular polyhedric structure and fine and medium pores. At depths over 60 cm, the skeleton is very abundant (>70%), consisting of pebbles and gravel.

Profile No. 1 represents soil class "C", with soils with a depth of about 100 cm, medium to silty texture, characterized by a surface layer of about 20 cm of dark-brown color, with 10–15% skeleton consisting of small and medium gravel; fine and medium polyhedric structure; and abundant fine pores. From 20 to about 700–800 mm deep, there is a brown/olive-brown layer with a frequent (25–35%) medium and large skeleton. It has medium and large angular polyhedric structures and fine pores. Over 700–800 mm deep the skeleton is abundant (35–50%), consisting of pebbles and gravel.

Two test plots of 10 × 20 m surface area were chosen (P1 and P3) as the plots to carry out deep burial stone reclamation digging the soil up to 1 m deep (Figure 4a). The dug soil was then sieved (Figure 4b), gradually discharging in the trenches the stones retained by the grid (Figure 4c); then, the filling of the trenches with the sieved fine earth, including skeletal fractions up to 40 mm (classes Y and Z) was completed (Figure 4d); levelling and compacting the layer of fine earth was performed with repeated passages of the crawler excavator (Figure 4e).

Figure 4. Diagram of the processing phase: (**a**) trenches digging; (**b**) soil sieving; (**c**) discharging of retained stones in the trenches; (**d**) refilling of trenches with sieved fine earth; (**e**) compaction of sieved fine earth layer.

Two other contiguous plots of the same area of undisturbed soil were delimited as control tests (P2 and P4) (Figure 5) [30].

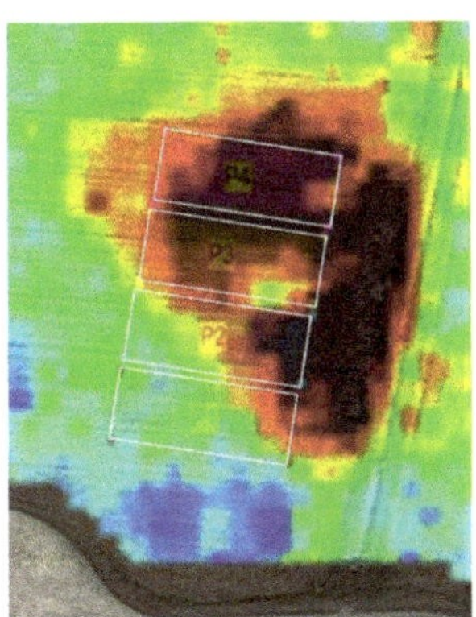

Figure 5. Experimental plots on soil geoelectric profiles (0–1 m deep): P1, P3 reclaimed plots; P2, P4, undisturbed control plots.

As a result, a clear improvement in the topsoil appearance was obtained without changing the plane of the field (Figure 6).

Figure 6. Appearance of the top soil on the P1 (reclaimed) vs P2 (undisturbed) experimental plots (40 m aerial picture).

2.2. Georadar Survey

The results of the described stone reclamation system on soil profile were evaluated by performing the Ground Penetrating Radar (GPR) mapping of the experimental plots using a GSSI UtilityScan RLT3 with 350 MHz HyperStacking antenna (Geophysical Survey Systems Inc., Nashua, NH, USA; Figure 7) [31].

(a) **(b)**

Figure 7. (**a**) GPR device used for the survey; (**b**) GSSI UtilityScan during the survey on experimental plots.

2.2.1. The Ground-Penetrating Radar Technology

GPR is a non-invasive geophysical technique [32] which employs radio waves, typically in the 10 to 3000 MHz frequency range, to map features buried in the ground. This nondestructive method detects the reflected signals from subsurface structures. When the signal hits a target or a boundary between materials having different permittivity values, it may be reflected from the surface and then detected by the GPR device. GPR is a convenient methodology for the imaging of subsurface features. It is an efficient investigation tool for studying the variations in the soil [33–35]. This technology can be used in agriculture to estimate the spatial pattern of lithological subsurface layers and reveal layer discontinuities [36]. This technique provides significant advantages over traditional methods such as soil cores or excavation.

2.2.2. Ground Penetrating Radar Survey

The GPR survey has proven able to detect the difference between two different contents of the soil. The survey was carried out in the 4 areas mentioned hereafter: A (P1, P3: reclaimed) and B (P2 and P4: undisturbed). The difference between A and B is related to the first shallow layer (60 cm depth). Soil A is more homogeneous than soil B because the stones have been removed and buried under the arable layer.

Twenty-two GPR profiles were collected inside the A and B areas by recording GPR traces (inline sampling) every 1 cm, with the maximum depth of investigation being 1.65 m (83 ns in terms of time window).

It has always been possible to detect the difference between soil A and soil B in the acquired profiles. The low backscatter of the signal is due to the lack of agglomerations and stones in the reclaimed areas (A) (Figure 8); meanwhile, in non-reclaimed areas (B), the signal has greater backscatter and attenuation along the depth axis (Figure 9).

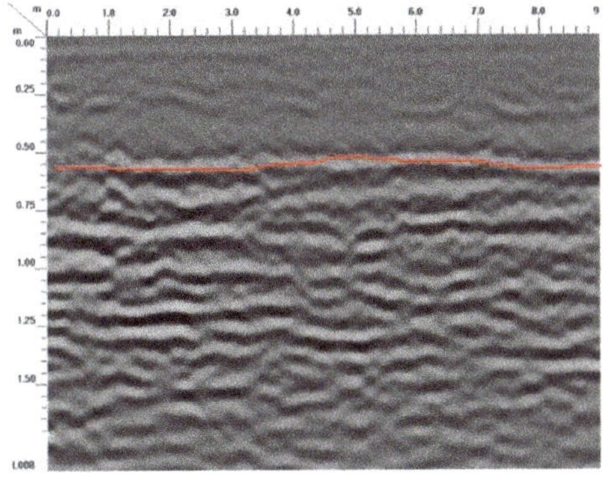

Figure 8. GPR profile collected on soil A (reclaimed soil). The red line shows the bottom layer of reclaimed soil.

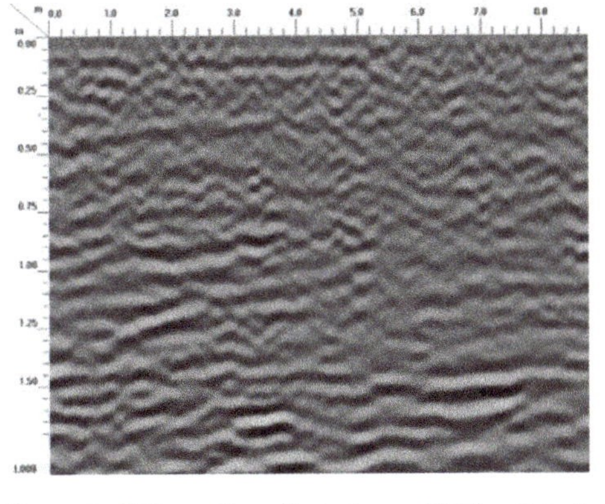

Figure 9. GPR profile collected on soil B (undisturbed) soil. The shallow layer is more scattered compared to the Figure 8 profile.

2.3. The Machines Used

On all plots, a preliminary seedbed preparation was carried out by harrowing with a combined tiller (Figure 10) at an approx. depth of 20 cm for the rupture of the soil crust and the burial of the previous crop residues.

Figure 10. Seedbed preparation harrowing.

The machines used were chosen to ensure the data record matched the chosen plots. For this reason, geo-positioning and auto-guidance systems were necessary to precisely ensure the three longitudinal passages.

The chosen tractor was a 117-kW nominal power tractor with continuously variable transmission (CVT) equipped provided with commercial PA technologies, consisting of a navigation system: GNSS receiver; real time kinematic (RTK) antenna, a terminal unit (VT), and a drive-wheel electric system for auto-guidance. The seeder was a 4.0 ISOBUS, pneumatic, 3 m wide, with 24 planter units, combined with disks (Figure 11).

Figure 11. The sowing tests setting.

2.4. Survey Methods

The considered factors are:

- Plot: 4, 2 reclaimed (B1; B2) and 2 undisturbed (NB1; NB2);
- Accelerometer position: 1 to 5;
- Axis of solicitation of the accelerometers: x; y; z.

The method involved three passes per plot at 1.25 ms^{-1} (4.5 km h^{-1}). The forward speed was established based on the experience of the operator as the maximum allowable in the worst-case plot scenario, and was kept constant in all the plots.

The acquired signals resulted in the three axes of the accelerometer (x = longitudinal; y = lateral; z = vertical) for all the 5 sensors and the microphone's sound pressure level. With this setting, the analyzed factors resulted in 16 channels. The frequency of acquisition was 1 kHz for the accelerometers and 50 kHz for the microphone.

Regarding noise, it will be reported as sound pressure level (SPL) and expressed in decibel (dB) unit. In order to consider the sound perceived by the human ear, the A-weighted decibel (dB_A) was adopted. Since it is assumed that the sound pressure levels may be higher than 85 dB_A, indicating a noisy environment, it is believed that the C-weighted decibel (dB_C) may be more suitable for the purpose of the study. For this reason, in addition to the commonly adopted dB_A, approach, the measurements will also be analyzed in dB_C. Regarding acceleration, the analysis included root mean square (RMS), maximum and minimum acceleration.

Consequently, the measurements made were:

- Accelerations: 3 (rep) × 5 (positions) × 4 (plots) × 3 (axis) × 3 (RMS, max, min) = 540;
- Noise: 3 (rep) × 4 (plots) × 2 (weighted dB) = 24 for a total of 564 values.

2.5. Instruments

The study focused on the measurement of acceleration and sound pressure.

The sensors used were accelerometers and microphones. The sensors, software and acquisition system are shown in Table 1.

Table 1. The instruments used.

Instrument/Material	Make/Model
Triaxial accelerometer	PCB 356A02
Triaxial accelerometer	Dytran 3263A2
Microphone	PCB 378B02
Data logger	Dewesoft Sirius
Data-acquisition software	Dewesoft X3
PC laptop	Panasonic CF31

The accelerometers used were all triaxial: three were positioned on the distribution elements of the machine, i.e., one on the left (position 1), one in the middle (position 2) and one on the right (position 3) (Figure 12a,b). Another was positioned on the main frame of the seeder (position 4), and the last one was placed on the pivoting roller (position 5).

The microphone was positioned in the horizontal position on the ladder railing and pointed toward the back of the machine.

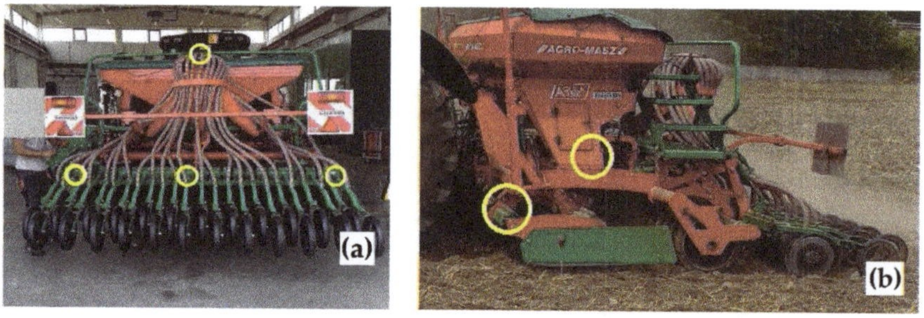

Figure 12. (**a**): the position of the triaxial accelerometers 1, 2 and 3 (on sowing elements, yellow circles) and the position of the microphone in the middle of the ladder railing of the upper portion on the hopper (yellow circle). (**b**): the position of the triaxial accelerometers 4 and 5 (yellow circles).

3. Data Analysis

The data acquired were processed with Minitab 17 statistical software [37] using the analysis of variance (ANOVA), which applied a generalized linear model (GLM) [38]. The plot (NB/B), the sensor position on the frame (1 to 5), and the acquisition axes (x, y, and z) were assumed to be fixed factors, while the root mean square (RMS) of the accelerations and the maximum and the minimum values of the time histories were the dependent variables. The Levene test ($p < 0.05$) [39] evaluated the homogeneity of the dataset's variances before the variance analysis.

4. Results

The results as root mean square (RMS) are reported in Table 2.

Table 2. RMS value of the acquired acceleration.

Plot	Accelerometer Position	Axis x (g)	Axis y (g)	Axis z (g)
B1	1	0.12	0.09	0.19
	2	0.09	0.13	0.17
	3	0.11	0.13	0.23
	4	0.06	0.08	0.14
	5	0.29	0.18	0.25
NB1	1	0.27	0.21	0.39
	2	0.18	0.25	0.34
	3	0.2	0.23	0.38
	4	0.1	0.13	0.22
	5	0.59	0.35	0.48
B2	1	0.14	0.14	0.24
	2	0.1	0.14	0.19
	3	0.1	0.11	0.22
	4	0.06	0.08	0.14
	5	0.3	0.2	0.26
NB2	1	0.23	0.21	0.37
	2	0.17	0.23	0.32
	3	0.19	0.24	0.38
	4	0.1	0.13	0.2
	5	0.47	0.31	0.41

The highest accelerations were found at the roller and ranged from 0.3 g in the destoned soil to 0.59 g in the undisturbed plots with a high skeleton level. The lowest accelerations, as expected, were at the frame, where the differences were less clear.

The mean acceleration value at the elements on the stony soils resulted in about double the destoned one.

Interestingly, the highest values were not found exclusively at the vertical axis, but all three axes were essential for the study. This aspect was observed, above all, at the pivoting roller (position 5), which appeared to be the most interesting: the longitudinal stress has the highest value (position 5; x axis).

The application of the analysis of variance (ANOVA) not only on the RMS values but also on the maximum and minimum values of the acquired time histories indicated the possibility of using acceleration to define the sowing parameters (Table 3).

Table 3. Results of the ANOVA.

Source	RMS 960904 *p*-Value	Max 960904 *p*-Value	Min 960904 *p*-Value
Plot	0.000 ***	0.000 ***	0.000 ***
Position	0.000 ***	0.000 ***	0.000 ***
Axis	0.000 ***	0.000 ***	0.035 *
Parcel x position	0.000 ***	0.000 ***	0.000 ***
Plot x axis	0.000 ***	0.005 **	0.529 ns
Position x axis	0.000 ***	0.000 ***	0.000 ***
Plot x position x axis	0.012 *	0.017 *	0.058 ns

*** = $p < 0.001$; ** = $p < 0.01$; * = $p < 0.05$; ns (not significant) = $p > 0.05$.

The plot, the sensor position, and the kind of sensor were all found to be significant factors ($p < 0.05$). Moreover, the data should be considered attentively as it is very interesting to note that their interaction was also significant. All the five accelerometers recorded significant differences. The accelerometer at the frame was the one with the smallest difference while that at the roller recorded the most differences. Regarding the accelerations of the distributors, measured as RMS, although these were the lightest elements, they were not those with the highest RMS acceleration, probably because they were fixed to the frame. As expected, the ANOVA analysis of the peak accelerations, both maximum and minimum, also gave significant results, confirming the different behaviors of the seeder when changing the plots. However, it must be noted that, in this case, it could be challenging to use this signal for monitoring in continuous mode and active optimization.

The results of the phonometric survey are reported in Table 4.

Table 4. Mean values of sound pressure level and result of the Tukey test.

Plot	Mean	Grouping	Plot	Mean	Grouping
	dBA			dBC	
B1	90.0	c	B1	102.1	b
NB1	98.7	a	NB1	110.5	a
B2	92.3	b	B2	104.1	b
NB2	98.2	a	NB2	110.2	a

Items with different letters are significant for $p < 0.05$.

In this case, too, a significant confirmed difference in noise value was found between the different plots. Both dB_A and dB_C gave significant results with very high differences (6 dB_A and dB_C). Therefore, both values could be adopted for monitoring the seeding. In this case, there is no signal problem as the software was already set up to provide the drive signal as both dB_A and dB_C. However, the results are of interest because it must be considered that noise is also the main feedback for the operator to decide the correct forward speed or the quality of the seeding.

5. Discussion

The question of the quality and speed of sowing is widely debated and current. The same sowing speed affects the outcome of the sowing. An increase in speed leads to a high degree of soil disturbance, inaccurate laying of the seed and suboptimal closure of the furrow. The issue was addressed by considering mainly two aspects: the properties of the soil and the technical and mechanical structure of the seeder.

In the literature, the issue is of particular interest in the fields of minimum tillage and soils with a high degree of stony disturbance. Many agricultural soils have variable percentages of coarse fractions in their arable layer, which interfere with crop growth and machine working. In addition, different techniques are available for destoning such as (i) collection and removal, (ii) on-site stone crushing, and (iii) stone burial. All these methods are expensive.

In addition, the further advent of geo-positioning, self-driving and isobus technologies have highlighted how the workability of the soil is a bottleneck for such machines. The limit highlighted by the operators is in the mechanical stress on the implement, which limits the operating speed to one third of the maximum possible.

With this aim, this study was performed to investigate whether it was possible to use noise and/or vibration values to monitor sowing performance in stony soils.

Four soil plots of two different disturbance degrees were prepared and an experimental sowing was carried out. The chosen forward speed of this experiment was 4.5 km h^{-1}, chosen as the maximum allowable speed on a stony plot by the operator.

Sowing trials on reclaimed and undisturbed plots showed that the noise and acceleration measured and acquired significantly differed when changing the soil conditions according to the stoniness degree of disturbance (DD) on soil workability.

Results confirmed that both noise and acceleration were significantly correlated with soil DD, to the point of being able to hypothesize their use as indicators to optimize mechanical operation.

Moreover, modern tractors with a global navigation satellite system (GNSS), auto guidance and ISOBUS seeder could also work at 15 km h^{-1}. This highlights not only the three-fold higher field efficiency of the machines but also the correct way of using the engine of high-powered tractors, whose power is limited by the maximum allowable speed. Therefore, the economic considerations about destoning could enable a new approach considering not only the current growing scarcity of further farming land, but also the opportunities offered by new digital technologies. For example, it is possible to imagine providing information on the correct forward sowing speed to the operator by means of acoustic or visual signals on the display or directly to the tractor's electronic control unit. The experiment also pointed out the capability of purpose-designed sensors (i.e., accelerometers) and their most interesting measurement axis. Moreover, use of the Tractor Implement Management technologies (TIM) system could optimize the operational efficiency of precision seeders in terms of working rate. TIM enables automatic operation, similarly to an on-the-go sensor, at the most suitable forward speed compatible with the structural characteristics of the machine depending on the levels of acceleration or noise induced in real time by the characteristics of the soil on which the machine operates, simultaneously reducing the risks of wear or breakage, and the related costs due to downtime and repairs. TIM's main concept is to enable bidirectional communication between tractor and implement, i.e., a transfer of control in both directions, to automatically control the main operational functions of the tractor.

Author Contributions: Conceptualization, P.T., M.C., E.P. and C.B.; methodology, P.T., M.C., E.P. and A.F.; software, M.B. and M.P.; validation, P.T., M.C. and M.B.; formal analysis, M.C., M.B., M.P. and N.C.; investigation, P.T., M.C., M.B. and M.P.; resources, M.C. and C.B.; data curation, P.T., M.C., M.B., A.F. and M.P.; writing—original draft preparation, P.T., M.C., M.B. and M.P.; writing—review and editing, P.T., M.C., M.B., M.P. and C.B.; supervision, P.T., M.C., M.B. and C.B.; funding acquisition, M.C. and C.B. All authors have read and agreed to the published version of the manuscript.

Funding: This research was funded by the Italian Ministry of Agriculture (MiPAAF) under the AgriDigit Programme, DDL n. 2111-B/2015.

Acknowledgments: The authors would like to thank, for their technical support: Gianluigi Rozzoni, and Ivan Carminati (CREA, Treviglio, Bergamo, Italy).

Conflicts of Interest: The authors declare no conflict of interest.

References

1. Colzani, G.; Cammilli, A.; Pirrone, S. Stato Di Pietrosità Dei Terreni E Lavorazioni Agricole State of Stoneiness of Soils and Agricultural Workability. *L'Informatore Agrar.* **1989**, *42*, 61–64.
2. He, D.; Lu, C.; Tong, Z.; Zhong, G.; Ma, X. Research Progress of Minimal Tillage Method and Machine in China. *AgriEngineering* **2021**, *3*, 633–647. [CrossRef]
3. Jiang, S.; Wang, Q.; Zhong, G.; Tong, Z.; Wang, X.; Xu, J. Brief Review of Minimum or No-Till Seeders in China. *AgriEngineering* **2021**, *3*, 605–621. [CrossRef]
4. Manquan, Z.; Yongwen, H.; Yueqin, L. Measurement and analysis on vibration characteristics of pneumatic seed metering device of no-till seeder. *Trans. Chin. Soc. Agric. Eng.* **2012**, *28* (Suppl. S2), 78–83. [CrossRef]
5. Xue, A.; Sheng, W.; Hongyan, D.; Chen, Y.; Yongchang, Y. Test on effect of the operating speed of maize-soybean interplanting seeders on performance of seeder-metering devices. *Procedia Eng.* **2017**, *174*, 353–359. [CrossRef]
6. Yang, L.; Yan, B.; Cui, T.; Yu, Y.; He, X.; Liu, Q.; Liang, Z.; Yin, X.; Zhang, D. Global Overview of Research Progress and Development of Precision Maize Planters. *Int. J. Agric. Biol. Eng.* **2016**, *9*, 9–26. Available online: https://ijabe.org/index.php/ijabe/article/view/2285/1179 (accessed on 1 July 2022).
7. Sharaby, N.; Doroshenko, A.; Butovchenko, A.; Legkonogih, A. A comparative analysis of precision seed planters. *E3S Web Conf. Innov. Technol. Environ. Sci. Educ.* **2019**, *135*, 1–19. [CrossRef]
8. Gasparetto, E.; Pessina, D. Seminatrici. Dispense del Corso di Meccanica Agraria (Lecture Notes of the Agricultural Mechanics course). Istituto di Ingegneria Agraria, Unimi. 1998. Available online: http://web.tiscali.it/profpessina/semina/seminatrici.htm (accessed on 1 July 2022).
9. MaterMacc Precision Planters. Available online: https://www.matermacc.it/eng/products (accessed on 1 July 2022).
10. *ISO 11783-1:2017*; Tractors and Machinery for Agriculture and Forestry—Serial Control and Communications Data Network—Part 1: General Standard for Mobile Data Communication. International Organization for Standardization: Geneva, Switzerland, 2017.
11. Maschio Gaspardo. Seminatrici di Precisione (Precision Sowing Machines). Available online: https://www.maschio.com/it/web/italia/seminatrici-di-precisione (accessed on 1 July 2022).
12. Staggenborg, S.A.; Taylor, R.K.; Maddux, L.D. Effect of Planter Speed and Seed Firmers on Corn Stand Establishment. *Appl. Eng. Agric.* **2004**, *20*, 573–580. [CrossRef]
13. Brandelero, E.M.; Adami, P.F.; Modolo, A.J.; Murilo Mesquita Baesso, M.M.; Adelar, J.F. Seeder Performance under Different Speeds and its Relation to Soybean Cultivars Yield. *J. Agron.* **2015**, *14*, 139–145. [CrossRef]
14. Liu, L.; Yang, H.; Ma, S. Experimental Study on Performance of Pneumatic Seeding System. *Int. J. Agric. Biol. Eng.* **2016**, *9*, 84–90. Available online: https://www.researchgate.net/publication/311439603 (accessed on 1 July 2022).
15. Nielsen, R.L. Planting Speed Effects on Stand Establishment and Grain Yield of Corn. *J. Prod. Agric.* **1995**, *8*, 391–393. [CrossRef]
16. European Commission: European Soil Data Center. Stony Soils Distribution in Western Europe. Available online: https://esdac.jrc.ec.europa.eu/public_path/Cfrag.png (accessed on 1 July 2022).
17. Flint, A.L.; Childs, S. Physical Properties of Rock Fragments and Their Effect on Available Water in Skeletal Soils. In *Erosion and Productivity of Soils Containing Rock Fragments*; Nichols, J.D., Brown, P.L., Grant, W.J., Eds.; John Wiley and Sons: Hoboken, NJ, USA, 1984; Volume 13. [CrossRef]
18. Badua, S.A.; Sharda, A.; Strasser, R.; Ciampitti, I. Ground speed and planter downforce influence on corn seed spacing and depth. *Precis. Agric.* **2021**, *22*, 1154–1170. [CrossRef]
19. Da Silveira, J.C.M.; Fernandes, H.C.; Modolo, A.J.; De Lima Silva, S.; Trogello, E. Furrow depth, soil disturbance area and draft force of a seeder-fertilizer at different seeding speeds. *Rev. Ceres Vicosa* **2011**, *58*, 293–298.
20. Barr, J.B.; Desbiolles, J.M.A.; Fielke, J.M. Minimising soil disturbance and reaction forces for high speed sowing using bentleg furrow openers. *Biosyst. Eng.* **2016**, *151*, 53–64.
21. Barr, J.B.; Desbiolles, J.M.A.; Fielke, J.M.; Ucgul, M. Development and field evaluation of a high-speed no till seeding system. *Soil Tillage Res.* **2019**, *194*, 104337. [CrossRef]
22. Colzani, G.; Cammilli, A.; Pirrone, S. Grado di disturbo alla lavorabilità dei terreni pietrosi (Degree of disturbance to the workability of stony soils). *L'Informatore Agrar.* **1989**, *44*, 39–43.
23. Pessina, D.; Facchinetti, D. Rateo variabile per una semina perfetta e sostenibile (Variable rate for perfect and sustainable sowing). *L'Informatore Agrar.* **2022**, *19*, 36–39.
24. Tiboni, M.; Remino, C.; Bussola, R.; Amici, C. A Review on Vibration-Based Condition Monitoring of Rotating Machinery. *Appl. Sci.* **2022**, *12*, 972. [CrossRef]

25. Mystkowski, A.; Kociszewski, R.; Kotowski, A.; Ci Ezkowski, M.; Wojtkowski, W.; Ostaszewski, M.; Kulesza, Z.; Wolniakowski, A.; Kraszewski, G.; Idzkowski, A. Design and Evaluation of Low-Cost Vibration-Based Machine Monitoring System for Hay Rotary Tedder. *Sensors* **2022**, *22*, 4072.
26. Yılmaz, D.; Gökduman, M.E. Development of a measurement system for noise and vibration of combine harvester. *Int. J. Agric. Biol. Eng.* **2020**, *13*, 104–108.
27. International Application Published under the Patent Cooperation Treaty (PCT). System and Method for Controlling Vibration on Agricultural Implement. WIPO Patent WO 2014/108722 A1, 17 July 2004.
28. IUSS Working Group WRB. *World Soil Resources Reports 2015, No. 106*; FAO: Rome, Italy, 2015.
29. Brambilla, M.; Romano, E.; Toscano, P.; Cutini, M.; Biocca, M.; Ferré, C.; Comolli, R.; Bisaglia, C. From Conventional to Precision Fertilization: A Case Study on the Transition for a Small-Medium Farm. *AgriEngineering* **2021**, *3*, 438–446. [CrossRef]
30. Toscano, P.; Cutini, M.; Cabassi, G.; Pricca, N.; Romano, E.; Bisaglia, C. Assessment of a Deep Burial Destoning System of Agrarian Soils Alternative to the Stone Removal and On-Site Crushing. *AgriEngineering* **2022**, *4*, 156–170. [CrossRef]
31. Codevintec. Available online: https://www.codevintec.it/media/strumentiPdf/gs-2004-utilityscan-eng-lo_KILux_it.pdf (accessed on 1 July 2022).
32. Daniels, D.J. *Ground Penetrating Radar, 2nd ed*; The Institution of Electrical Engineers: London, UK, 2004.
33. Davis, J.L.; Annan, A.P. Ground Penetrating Radar for high resolution mapping of soil and rock stratigraphy. *Geophys. Prosp.* **1989**, *37*, 531–551.
34. Lombardi, F.; Lualdi, M. Step-Frequency Ground Penetrating Radar for Agricultural Soil Morphology Characterization. *Remote Sens.* **2019**, *11*, 1075.
35. Zhang, J.; Lin, H.; Doolittle, J. Soil layering and preferential flow impacts on seasonal changes of GPR signals in two contrasting soils. *Geoderma* **2014**, *213*, 560–569.
36. Allred, B.; Daniels, J.J.; Ehsani, M.R. *Handbook of Agricultural Geophysics, 1st ed*; CRC Press: Boca Raton, FL, USA, 2008.
37. Minitab 17. Statistical Software. State College, Pa.: Minitab, Inc. 2010. Available online: www.minitab.com (accessed on 4 July 2022).
38. Hastie, T.J.; Pregibon, D. Generalized linear models. In *Statistical Models in S*; Chambers, J.M., Hastie, T.J., Eds.; Wadsworth & Brooks/Cole: Pacific Grove, CA, USA, 1992; Chapter 6.
39. Levene, H. *Contributions to Probability and Statistics: Essays in Honor of Harold Hotelling*; Stanford University Press: Redwood City, CA, USA, 1960; pp. 278–292.

AgriEngineering

Article

A Full Assistance System (FAS) for the Safe Use of the Tractor's Foldable Rollover Protective Structure (FROPS)

Davide Gattamelata [1], Daniele Puri [1], Leonardo Vita [1] and Mario Fargnoli [2,*]

[1] Italian Workers' Compensation Authority (INAIL), Monte Porzio Catone, 00078 Rome, Italy
[2] Engineering Department, Universitas Mercatorum, 00186 Rome, Italy
* Correspondence: mario.fargnoli@unimercatorum.it

Abstract: The use of agricultural tractors is a major concern in agriculture safety due to the high level of risk of loss of stability combined with the frequent absence of passive safety devices such as rollover protective structures (ROPSs). Indeed, although in most cases the ROPS is installed, when working in vineyards, orchards, or in other cases of limited crop height, the tractor is usually equipped with a foldable ROPS (FROPS), which is often misused because the effort needed for raising/lowering is excessive and the locking procedure is time-consuming. Thus, the goal of this research is to investigate the problem from the ergonomics point of view, developing a support system capable of facilitating FROPS operations. The research outcome consists of the development of a retrofitted full assistance system (FAS) for lowering/raising the FROPS by means of electric actuators. Additionally, an automatic locking device (ALD) was also developed to safely and automatically lock the FROPS. Both the FAS and ALD systems were implemented following a reverse-engineering approach, while their final validation was performed by means of a real prototype tested in a laboratory. The results achieved can contribute to expanding knowledge on human-centered research to improve safety in agriculture and thus social issues of sustainable agricultural systems.

Keywords: agriculture; machinery safety; occupational health and safety (OHS); ergonomics; agricultural tractors; rollover protective structure (ROPS); reverse engineering; virtual prototyping; sustainable agricultural systems

Citation: Gattamelata, D.; Puri, D.; Vita, L.; Fargnoli, M. A Full Assistance System (FAS) for the Safe Use of the Tractor's Foldable Rollover Protective Structure (FROPS). *AgriEngineering* **2023**, *5*, 218–235. https://doi.org/10.3390/agriengineering5010015

Academic Editors: Marcello Biocca and Roberto Fanigliulo

Received: 16 December 2022
Revised: 21 January 2023
Accepted: 23 January 2023
Published: 25 January 2023

1. Introduction

Putting them into practice, safety issues represent the operational command of sustainability [1], and occupational health and safety (OHS) initiatives aimed at reducing accidents represent the operationalization of sustainability in workplaces [2,3]. Accordingly, a human-centered approach to improve safety is regarded as a means to improve sustainable working systems [4]. In agricultural activities, although a decrease in accidents was registered in the last years, mainly due to the reduction in activities related to the COVID-19 pandemic, the number of injuries and fatalities is still great if compared to other sectors [5], as well as the occurrence of health diseases such as musculoskeletal disorders (MSDs) [6], and most severe injuries and fatalities concern the misuse of tractors [7]. In particular, the tractor's rollover is reported as the main cause of fatalities [8,9], especially when the rollover protective structure (ROPS) had not been installed (e.g., in old vehicles) or when it was not correctly used. Actually, the ROPS is a roll bar that absorbs energy in the case of the tractor's rollover, enabling a safety zone for the operator, which is called a "clearance zone".

In Figure 1, two examples of two-post foldable ROPSs (FROPSs) for narrow-track tractors are shown: a rear-mounted FROPS on the left and a front-mounted FROPS on the right side of the figure.

Figure 1. Examples of a rear-mounted FROPS (**left**) and a front-mounted FROPS (**right**).

To reduce such a phenomenon, different initiatives have been carried out worldwide [10], such as the updating of old tractors with ROPSs and seatbelts, which has been promoted in different countries by means of retrofitting campaigns [11–13], the issuing of technical guidelines [14,15], or training courses for farmers [16].

However, these initiatives hardly impact the misuse of foldable ROPSs; for example, when the foldable ROPS is disabled to better operate in vineyards, orchards, or in other cases of limited crop height and is not subsequently raised after this work. These situations are very common, as reported by numerous studies showing that up to 50% of fatalities are due to the misuse of foldable ROPSs [17–19].

According to occupational health and safety (OHS) guidelines, farmers can unfold the ROPS in cultivation work where there is a low clearance between the tractor and the vegetation (both in height and in width), but they should raise it again as soon as this specific work is completed. Instead, when operating in these environments, farmers usually keep it unfolded; such an unsafe behavior is very common and it is typically justified by farmers pointing out the excessive effort that they have to make to fold/unfold the ROPS several times in a work day [20,21]. It must be noted that the tractors used most in these operations are narrow-track models, which can be defined per Molari and Rondelli [22] as a tractor that, when equipped with tires of the greatest allowed dimensions, "has a minimum track width of not more than 1150 mm".

The ergonomics literature has provided different approaches to deal with this problem: for example, Cremasco et al. [23] developed and tested a prototype solution to increase FROPS reachability, which consists of a rod that allows the users to raise a two-post rear-mounted FROPS more easily, respecting anthropometric variability. Differently, Etherton et al. [24] developed a telescoping structure for an automatically deployable ROPS. Similarly, Alkhaledia et al. [25] proposed an automatically foldable protective structure, which resulted in a very effective solution to protect the operator. The development of automatic solutions was investigated also by Ballesteros et al. [19], who proposed an automatically deployable (both in height and width) ROPS. All these mechanical solutions are noteworthy from the technical point of view, but often they are inconvenient from the financial one, especially in the case of upgrading old tractors. Moreover, they scarcely analyze the ergonomic issues, which leads to discomfort for the farmer when operating the ROPS. As remarked by recent studies [20,26] on this topic, further research is needed to make the ROPS's lowering/raising more comfortable. Such an address was followed by Gattamelata et al. [27], who discussed the development of a partial assistance system

(PAS) to reduce the operator's physical effort: on the one hand, this system can improve the usability of the FROPS because it reduces the loads in lowering/raising operations by means of springs; on the other hand, such a solution represents a partial solution to reduce the operator's discomfort and further analysis is needed to optimize the configuration and implement a safe locking system to practically increase the FROPS's usability. In fact, although decreasing the operator's stress in lowering/raising operations, such a solution does not avoid manual handling, which is the most common cause of FROPS misuse.

Hence, the goal of the current study is to expand the research outputs of Gattamelata et al. [27], focusing on the following issues:

- The development of a full assistance system (FAS) by means of strength analysis;
- The development of a system that can automatically lock the FROPS when it is unfolded without making the folding operations strenuous;
- The constructive and dynamic integration of this automatic locking device (ALD) with the FAS;
- The physical testing of the prototype.

The remainder of the article is organized as follows: in Section 2, the research approach is summarized, starting with the preliminary analysis related to the FROPS's lowering/raising operations, which is illustrated in Section 3, providing the input criteria used for the design activities. In Section 4, the concrete experience analysis is summarized. Then, in Section 5, the design of the FAS is proposed, while in Section 6, the development of the ALD is addressed. Section 7 shows the results of both the virtual and practical prototyping of these systems. Section 8 discusses the outcomes of the study, while Section 9 contains conclusive remarks.

2. Research Approach

The development of the FAS and ALD system was carried out following a reverse-engineering approach, combining design for safety and ergonomics tools with a bottom-up approach [28]. As underlined by Wood et al. [29], such an approach mainly relies on the following activities:

- Predicting the system behavior;
- Predicting human behavior;
- Analyzing the system functions;
- Analyzing the users' behavior;
- Developing solution principles and the associated functions of the system;
- Developing a virtual model and validating it;
- Developing a physical prototype and validating it.

More in detail, the flow of activities that should be carried out can be schematized as in Figure 2. Needless to say, for each activity, the tools used to perform it can vary depending on the case context and engineers' needs.

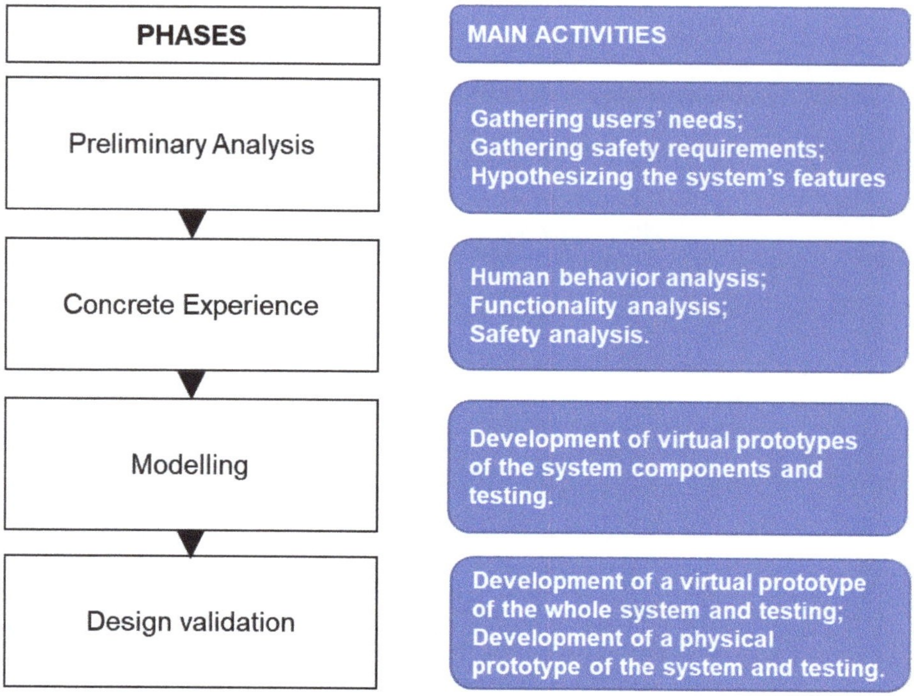

Figure 2. Scheme of the proposed research approach.

3. Preliminary Analysis

As previously mentioned, the most common tractor types equipped with foldable ROPS are those that can operate where there is a low clearance between the tractor and the vegetation, i.e., the narrow-track tractors, whose main features based on the OECD Codes are the following [30]:

- A ground clearance not higher than 600 mm considering the lowest points of the axles;
- A minimum track width with one of the axles less than 1150 mm when the tractor is equipped with tires or tracks of the largest size recommended by the manufacturer;
- The unladen mass of the tractor, which can vary from 400 kg up to 3500 kg.

Moreover, the OECD Codes provide information on the ergonomic features of the foldable ROPS such as the criteria that must be used to measure loads in manual raising and lowering operations [27]. In such a context, useful definitions to determine the geometrical characteristics of these operations are:

- "Grasping area", which is the part of the FROPS used by the operator to raise/lower the bar;
- "Accessible part of the grasping area", which represents the area that can be reached by the operator when raising/lowering the FROPS;
- "Accessible zone", consisting of the volume occupied by a standing operator when raising/lowering the FROPS.

Accordingly, the OECD Codes [30] indicate the acceptable force limits for raising/lowering operations, which can vary based on the different accessible zones: with reference to Figure 3, the acceptable force limit is:

- For zone I, 100 N;
- For zone II, 75 N;
- For zone III, 50 N.

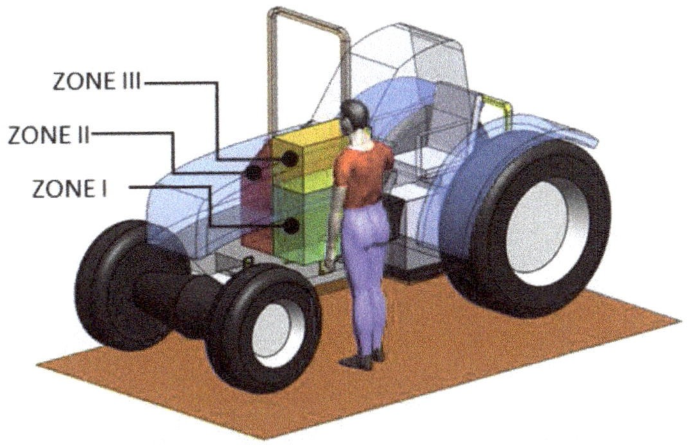

Figure 3. Accessible zones of the grasping area for a wheeled tractor according to OECD Code 6 criteria.

These values can also vary based on the different operations: for example, an increase in the force limits of up to 50% is allowed for lowering operations, and when the FROPS is fully raised or fully lowered, the force limit can be augmented by up to 25%. However, these indications are not sufficient at a practical level due to the variability of the operators and their real positioning when grasping the FROPS [23], which increases the efforts needed to operate the bar, leading to its misuse when working in fields. These criticalities have been brought forward by different researchers [31–33], who agree on the need to develop more human-centered solutions to reduce FROPS misuse. For the purpose of this study, a narrow-track tracklaying tractor equipped with a front-mounted two-post FROPS was used (Figure 4). It must be noted that in this figure, α represents the folding angle, and the rest configuration of the FROPS (folded FROPS) does not correspond to $\alpha = 0$ because of the constructive features of the tractor's bonnet. Actually, such features are very common for tracklaying tractors equipped with front-mounted FROPSs.

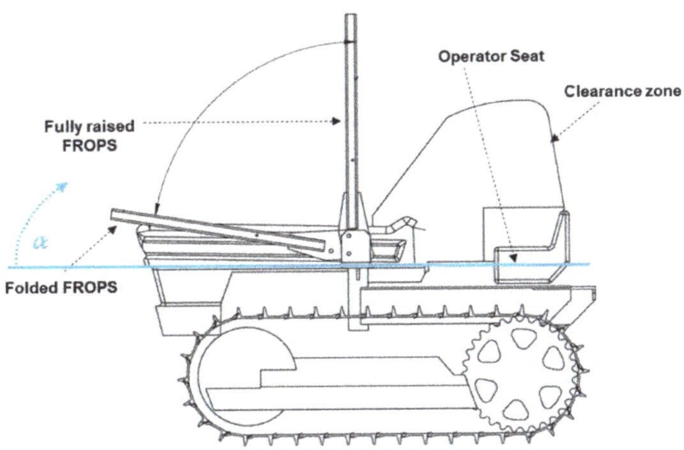

Figure 4. Scheme of the analyzed tractor (α represents the folding angle of the FROPS).

4. Concrete Experience

Following the proposed approach, a practical analysis of the ergonomics features of the FROPS' use was carried out involving five tractor users to better define the dynamics and geometrical characteristics of raising/lowering operations practically. Each operator was asked to fold/unfold the FROPS several times, allowing us to register the body position and the handling points.

From the analysis of the raising/lowering operations, it emerged that the effort needed by the user to overwhelm the weight force of the FROPS when unfolding it and to maintain the bar when folding it is higher than the force limits foreseen by the OECD Codes. These results are in line with previous studies [31,34], showing that the risk of misuse of the FROPS is very high on the one hand, and the risk of musculoskeletal diseases should also be taken into account on the other, since raising/lowering operations might be repeated several times daily. Actually, it has to be considered that the weight of the FROPS is about 72 kg, while the roll bar height reaches 1210 mm, as per Gattamelata et al. [27].

Moreover, another issue that is scarcely treated in the extant literature is related to the risks that the operator is exposed to if the FROPS locking/unlocking operations are not performed correctly. In fact, to block the FROPS, two pins have to be used by the operator, who needs to walk around the tractor and fix them on both sides. These operations can contribute to FROPS misuse since additional physical efforts and a waste of time from the production point of view are required. Furthermore, the risk of entanglement and the risk deriving from the incorrect locking of the roll bar should also be considered.

Based on these considerations, the use of a full assistance system (FAS) that can support the operator in the FROPS raising operations, on the one hand, and the inclusion of an automatic locking device (ALD) to block the bar when fully raised, on the other, can certainly reduce the above-mentioned risks. The development of these safety devices was carried out considering their applicability to both new vehicles and old ones, i.e., the retrofitting of tractors already in use.

5. Modeling the Full Assistance System

The selection of the model was made considering the most diffused components and mechanical systems available on the market: it was found that the assistance system can be a linear or a rotating actuator, which acts on the FROPS, supporting its rotation with respect to the hinge joint. We decided to implement a linear actuator, as illustrated in Figure 5.

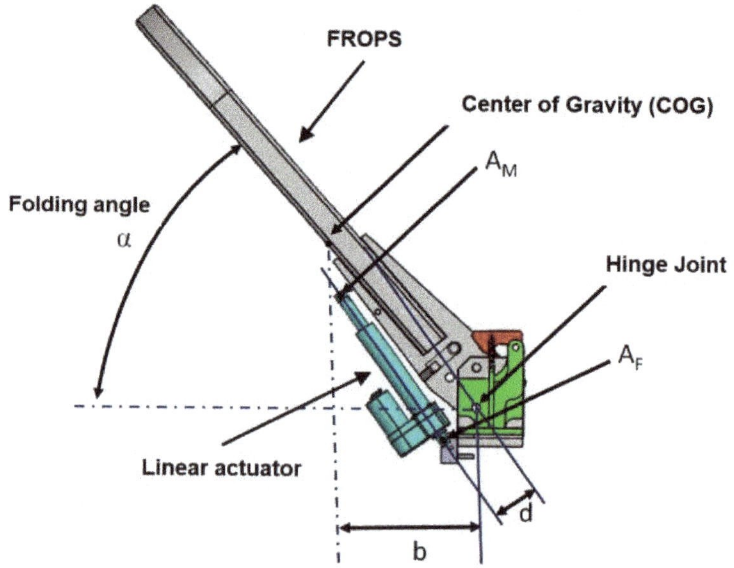

Figure 5. Model of the geometrical features of the linear actuator (A_F and A_M represent the anchorage points of the actuator).

This choice is motivated by the fact that the FAS we are developing should be destined for the ROPS retrofit as well, limiting the costs that farmers have to bear for this type of upgrade. Accordingly, the electric linear actuator can be easily supplied by the tractor battery, and for this reason, it is much cheaper than a hydraulic or pneumatic system.

The dynamic analysis of the FROPS was based on the results of the ergonomic study proposed by Gattamelata et al. [27] and was aimed at evaluating the forces involved in the operation of folding/raising by means of the linear actuator. More in detail, the following moments were defined:

- M_W is the moment of the weight force of the FROPS: this moment varies depending on the horizontal distance of the center of gravity (GOC) from the axis of the hinge joint;
- M_{ACT} represents the moment of the actuator, which varies on the basis of both the actuator type and the forces it can exert on the FROPS, considering the distance between the anchorage points A_F (fixed anchorage point) and A_M (mobile anchorage point) on the one hand, and the hinge axis to determine the arm lever on the other.

Accordingly, during the lowering/raising phase, M_{ACT} should be greater than the moment of the weight force, and the expression (1) must be satisfied:

$$M_{ACT} \geq M_W \tag{1}$$

The moment of the weight force M_W can be determined by means of Equation (2) considering the folding angle α:

$$Mw = P \times b \times \cos(\alpha) \tag{2}$$

- where P is the weight force of the FROPS, b is the arm of the weight force, and α represents the inclination of the FROPS (i.e., the folding angle), as schematized in Figure 5. It must be noted that M_W reaches its maximum value in the rest configuration, and based on the data provided by [27], this value is about 187 Nm for each side.
- The design of a suitable FAS requires an iterative procedure consisting of three main steps:
- Step 1—identification of the actuator model in terms of strength (F_{ACT}) and stroke (i.e., the maximum and minimum distance between A_F and A_M allowed by the system);
- Step 2—definition of the A_F point on the FROPS's fixed part and of A_M on the FROPS's mobile part;
- Step 3—analysis of the load capability condition ($M_{ACT}(\alpha) \geq M_W$).

It is noteworthy that in the second step, to allow an easy retrofit, a graphical procedure can be used that mimics the actuator behavior to find the position of the fixed and mobile hinge points without complex geometric calculations. Such a procedure is summarized in Figure 6 through three phases:

- Phase 1: Evaluation of the angular range of motion of the FROPS, i.e., the span between the rest configuration (FROPS folded down) and safety configuration (FROPS fully raised); definition of the point A_M;
- Phase 2: Definition of the position of A_M in the rest configuration (A'_M) considering the sector of the circumference with the center corresponding to the projection of the hinge joint O (the amplitude corresponds to the angle $\Delta\alpha$);
- Phase 3: Point A_F of the actuator can be obtained as the intersection between the two circular arcs r and s, where s is the circumference arc centered at point A_M (radius S), while r is the circumference arc centered in A'_M (radius R). The anchorage points comply with the actuator range of motion only if R is smaller than the maximum elongation of the actuator L_{max} and S is smaller than the minimum elongation of the actuator L_{min}. If the actuator length and elongation do not meet these geometrical requirements, the A_M point or/and the A_F point should be changed iteratively.

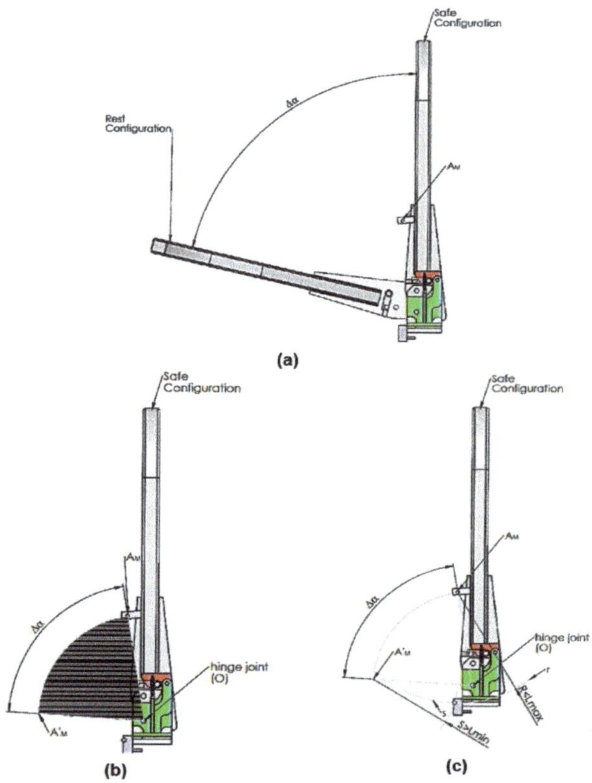

Figure 6. Procedure to define the A_F and A_M points: (**a**) determination of the angular range of motion $\Delta\alpha$ and A_M; (**b**) determination of the A_M point in the rest configuration (A'_M); (**c**) determination of the fixed anchorage point of the linear actuator (A_F).

Finally, it must be observed that step 3 concerns the dynamic validation of the actuator: if Equation (1) is not satisfied, it is necessary to restart the procedure with another actuator model. In this study, considering the inertia and dimensions of the prototype, and to be in compliance with Equation (1), an electric actuator with the following characteristics was chosen:

- Nominal stroke: 152.4 mm;
- Voltage: 12 V DC;
- Current draw: 20 Amp;
- Speed: 33 mm/sec;
- Load: up to 18,000 N.

6. Modeling the Automatic Locking Device

The presence of a full assistance system to raise the FROPS eliminates the need to manually operate it, but in order to ensure the operator's safety in case of a rollover, the FROPS must be locked in a safe configuration. The locking phase is usually made by the operator, who has to insert the pins in the locking holes on both sides of the tractor. The implementation of an automatic locking device (ALD) can avoid these operations and the risk of misusing the locking device. To achieve such a goal, it should be taken into account that an ALD must guarantee both safe locking and suitable structural resistance. These requirements, together with the need to keep the costs of the device at a feasible level, led us to implement a mechanical solution, capable of locking the roll bar automatically at the end of the raising phase and withstanding the loads deriving from the potential impact with the ground in case of a rollover. For these reasons, the ALD is preliminarily virtually prototyped to perform kinematic, dynamic, and strength analyses. Once the virtual model of the ALD satisfies all the performance requirements, a physical prototype will be developed and installed on a FROPS in order to carry out experimental tests of the whole system in accordance with the OECD Codes. In Figure 7, the model of this device assembled with the FAS is shown, describing its functionality that consists of three main phases:

- Triggering;
- Engagement;
- Locking.

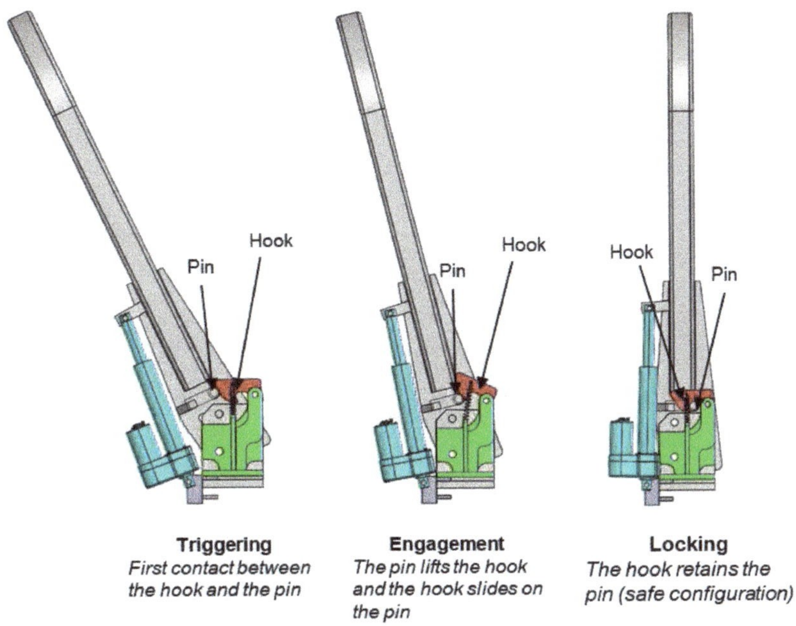

Figure 7. Operating phases of the locking device.

7. Validation

7.1. Virtual Testing

Once the geometrical features are defined, the whole system integrating both the FAS and the ALD was developed virtually to test its kinematical and dynamic features. Actually, dynamic analyses are useful to evaluate the correct engagement of the automatic locking device (i.e., the correct coupling between the profiles of the hook and the circular profile of the locking pin) and the effect of the FROPS inertia on the performance of the linear actuator. These issues were investigated by means of a multibody dynamics simulation software system (MSC Adams [35]).

Firstly, the correct coupling of the hook with respect to the pin on the FROPS was evaluated. In detail, the virtual testing consisted of verifying the hook profiles to optimize the engagement configuration and the locking one, as shown in Figure 8. In other words, the hook must have the two following features: on the one hand, the external profile has to be hit by the pin and raised without excessive effort by the actuator; on the other hand, the internal profile should constrain the pin as long as the hook is deliberately disengaged.

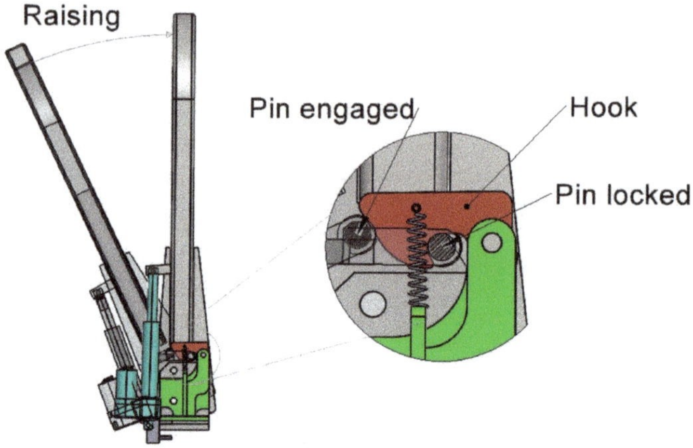

Figure 8. Scheme of the engagement and locking configuration.

For the external profile of the hook, the most important geometrical feature is the inclination: thus, by simulating different values of the inclination angle of the external profile, the one requiring less dynamic variation force on the actuator during the hitting and sliding of the pin with the hook was chosen. This characteristic is well highlighted in Figure 9, where a post-processing multibody simulation diagram of the actuator force with respect to the raising angle is shown. It must be noted that the raising angle α' is obtained by Equation (3) to take into account the fact that the rest configuration of the FROPS does not start at $\alpha = 0$ (see Figure 4):

$$\alpha' = \alpha - 6^\circ \quad (3)$$

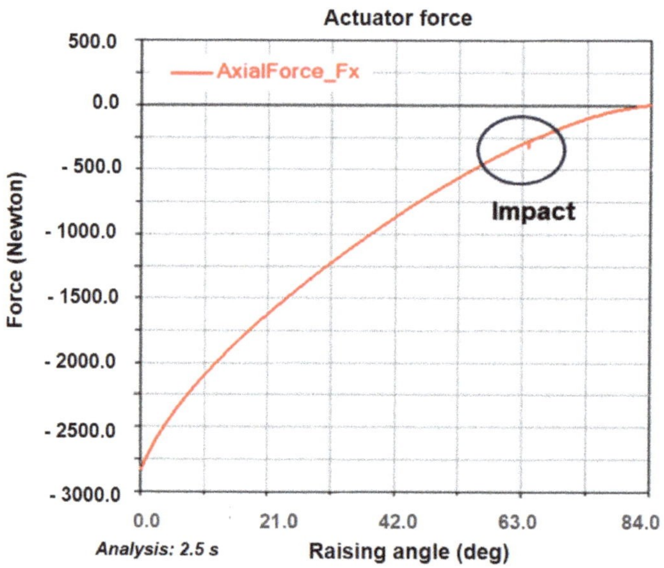

Figure 9. Dynamic simulation of the system (raising time 2.5 s and constant speed of the actuator).

It is noteworthy to mention that the diagram has a little jitter at 63.5 degrees when the pin comes into contact with the hook and raises it.

In Figure 10, the 3D model used for this simulation is shown.

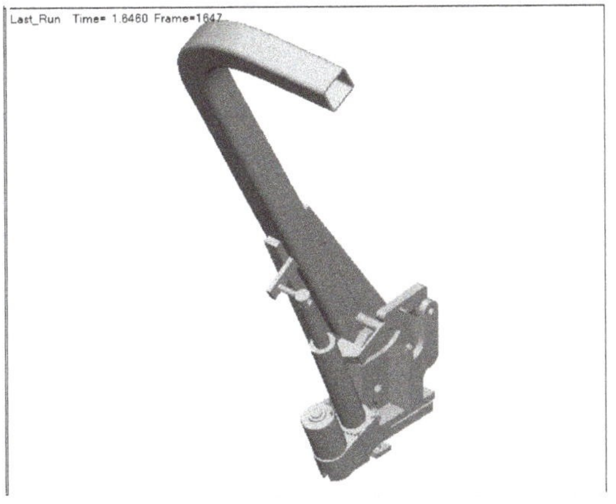

Figure 10. Three-dimensional model of the system used for the dynamic simulation.

Furthermore, additional simulations with different raising times were performed to investigate the dynamic effect of the FROPS's raising speed on the impact of the pin with the hook. Although numerous simulations were carried out, in Figure 11, the force behavior for a constant speed of the actuator corresponding to the following three raising times is shown:

- One second, which is in line with the time of the tractor's rollover;
- Five seconds, which corresponds to the time needed by the chosen actuators to fully raise the FROPS;
- Two and a half seconds, which is an intermediate value between 1 and 5 s that can add information on the effects of the impact on the ALD.

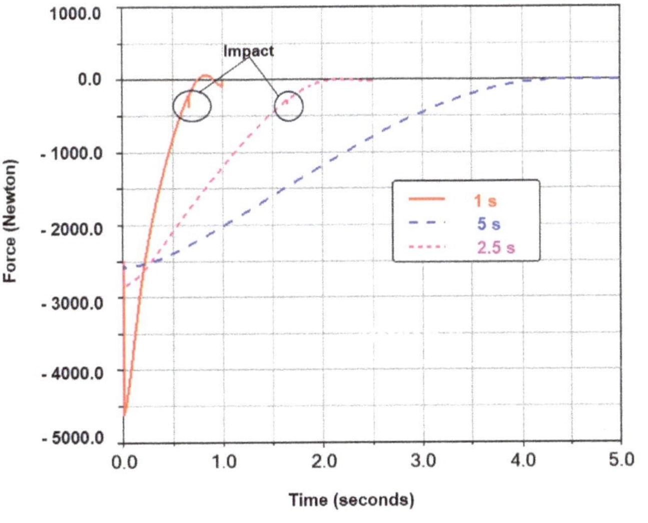

Figure 11. Comparison of the FAS raising force when adopting three raising times (1 s, 2.5 s, and 5 s).

The impact of the pin and the hook has dynamic effects on the actuator only for a raising time of less than 2.5 s, while for a 5 s raising time, there is no effect on the actuator since the maneuver is slow. The raising time variation in the multibody dynamic simulation made it possible to quantify the outcome of the FROPS's inertia when operating it with

FAS and ALD systems. As shown in Figure 11, a shorter lifting time requires a stronger as well as faster actuator. Actually, a raising time of about 2.5 s requires an input force of the actuator of about 2840 N. In contrast, the 5 s raising time requires a force of 2585 N, while reducing the raising time to 1 s means adopting an actuator with a 4622 N input force at least, which is about 62.7% greater than the one required for the 2.5 s raising time, which is the selected option.

For the validation of the internal profile of the hook, the most important feature is the curvature, which must be greater than the pin radius value, but it cannot be too large since the hook must hold the pin in case of rear loading on the FROPS (i.e., when the tractor overturns backward due to a wheelie). For this purpose, the structural analysis of the hook allowed us to analyze if the locking device can bear the expected loads and retain the roll bar at the same time. The strength of the system was tested in accordance with the criteria provided by the OECD codes [30]. More in detail, the proposed locking device with hook and pin was tested with respect to the longitudinal load by means of the finite elements method (FEM). First, we considered the loads reached by a traditional structure characterized by manual locking pins: in this case, a load of about 21,000 N was applied at the upper part of the roll bar, considering that:

- According to the authors' experience in testing retrofitted ROPSs [7,13,27], a load of about 21,000 N is likely to meet the test criteria required by the OECD Codes and it was applied at the upper part of the roll bar;
- The roll bar and plate material is S275 JR steel;
- The locking pins were obtained from calibrated bolts of 10.9 grade (yield strength 940 MPa).

The contour plot of the Von Mises stress is shown in Figure 12, where the stress values are expressed in MPa.

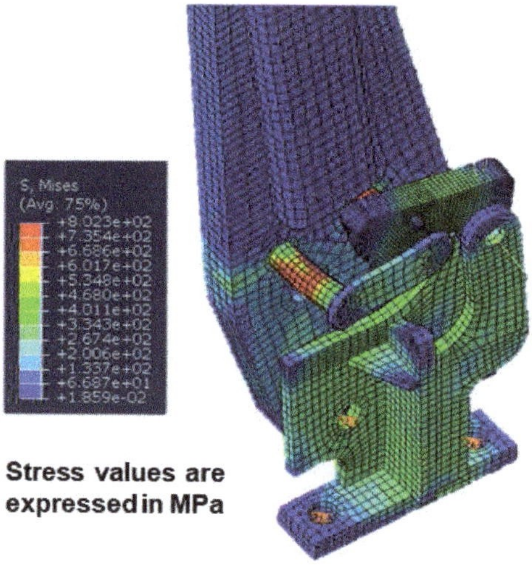

Figure 12. Von Mises stress diagram of the longitudinal load (energy value 4500 Joules).

It is important to underline that the developed FEM analysis represents a preliminary investigation to check if the ALD is able to withstand the loads that are hypothetically faced by the FROPS during a physical test. Then, it is always necessary to perform physical tests in order to assess the capability of the whole structure to act as a ROPS according to the OECD Codes.

7.2. Physical Testing

Physical testing at the moment has been completed only for the validation of the ALD design, while the FAS system is still being analyzed.

In detail, a real prototype of the FROPS equipped with the locking system was realized and tested by means of the ROPS test rig set in the INAIL (the Italian Workers' Compensation Authority) research center located in Monte Porzio Catone (Rome). The equipment installed in the laboratory can allow engineers to test the ROPS in accordance with the OECD Codes, thus ensuring the compliance of the system with current safety requirements. Due to privacy concerns, the description of the results achieved is simplified.

In Figure 13, the preliminary setting operations are shown, while in Figure 14, the test of the FROPS equipped with the ALD is shown, where a longitudinal force is applied at the top of the ROPS, pushing it from the front to the rear.

Figure 13. Preliminary setting operations.

Figure 14. Application of a longitudinal force to test the behavior of the ALD.

Indeed, this longitudinal force represents the most severe condition for the locking system, and the energy required by the OECD Code 7 [30] for a narrow-track wheeled tractor having a mass of 2000 kg was chosen as a benchmark. In practice, during the test, a maximum force of 20,480 N was reached, corresponding to the maximum deformation of the FROPS in the direction of the force, which was 223 mm (Figure 15).

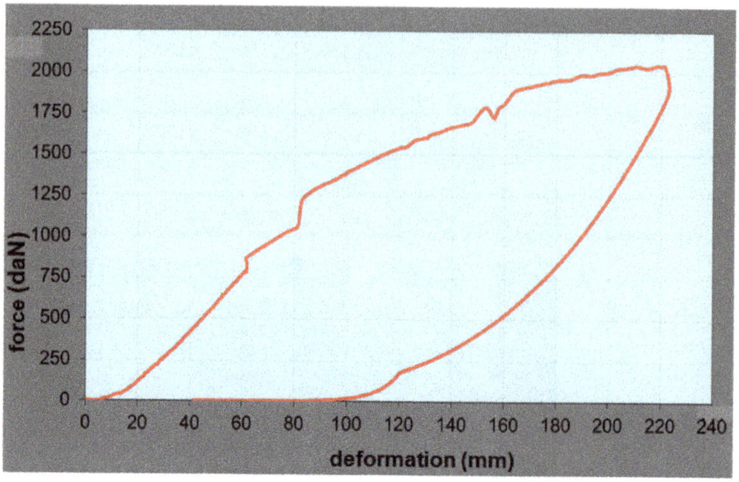

Figure 15. The plot of the plastic deformation.

The residual plastic deformation registered was about 96 mm. During the test that simulated the hit of the ground in case of a rollover, the locking system was able to maintain the FROPS in the upright locked configuration (safe configuration) and no significant deformations of the ALD were registered. After the removal of the force, both the locking systems (left and right) were still capable of locking the FROPS in the upright position as well as unlocking it for folding. Thus, the results achieved in terms of maximum deformation, plastic deformation, and maximum load (necessary for evaluating the energy absorbed by the structure) are in line with the results of tests developed on a similar FROPS with locking pins and mountings [36].

8. Discussion

Human interaction with mechanical systems represents a key factor in occupational safety in many sectors [37,38] and it is the cause of most severe injuries and fatalities occurring in agricultural activities [39]. Irwin and Poots [40] as well as Caffaro et al. [41], to cite a few, underlined the relevance of fatigue, time pressure, and stress as the main factors that can lead to the misuse of machinery among farmers Accordingly, a human-centered approach is needed to develop technical solutions aimed at facilitating working activities and reducing the risk of the misuse of work equipment such as tractors [42–44].

The current study represents a practical answer to these research hints through the development of a full assistance system (FAS) that can be applied to two-post FROPSs, which are usually equipped on narrow-track tractors. In addition, the FAS's implementation was integrated with the development of a specific automatic locking device (ALD). The use of these combined systems can certainly reduce the risk of FROPS misuse since it eliminates manual handling by the operator. Moreover, the full assistance device also eliminates the risks of musculoskeletal diseases and those related to maneuvering the bar (whose weight is more than 70 kg in the analyzed case study), such as entanglement, cuts, and crashing.

The reliability of both the FAS and ALD has been verified by means of virtual modeling and testing: the output of these analyses demonstrated that the proposed approach can be suitable for retrofitting two-post FROPSs for narrow-track tractors. Indeed, considering the features of the components analyzed, the developed solutions can represent a useful reference for the implementation of FAS and ALD systems to a large variety of tractors already in use. Hence, these outcomes can be used to effectively increase the spread of retrofitted safety solutions for agricultural tractors, consistent with Kogler et al. [45], who stressed the need to augment information concerning safety solutions that can reduce the occupational risks of agricultural tractor users.

Compared to the development of similar technical solutions, such as the AutoROPS by Etherton et al. [24] or similar attempts [19,25], the FAS and ALD combined system presented in the current study presents the following advantages:

- The system can be easily adapted to existing FROPS models, even if, in this case, an additional structural test is necessary to verify the compliance of the modified model of the protective structure with the OECD Codes;
- The implementation costs are very low considering that the two electric actuators (one for each side of the FROPS) can be connected to the tractor's battery for the energy supply;
- Lowering/raising and locking operations can be carried out by the operator from the driving seat.

From the methodical point of view, the proposed approach is in line with the research findings of Casazza et al. [46], providing a detailed description of a reverse-engineering procedure that can be followed by researchers and practitioners to deal with the development of similar devices. Indeed, such an approach is effective when technical components have to be developed to upgrade existing machinery, consistent with Urbanic and El Maraghy [47]. Thus, FAS and ALD systems can be developed to equip both new tractor models and existing ones that should be retrofitted. These criteria can make the updating of tractors easier, facilitating compliance with OHS requirements [48].

Besides these positive results, the limitations of the study also need to be underlined. First, the application of the proposed approach requires a more extended validation concerning:

- The physical tests of the FROPS equipped with the ALD to verify compliance with the OECD Codes;
- The completion of the practical implementation of both systems on the tractor.

The integration of the FAS and ALD systems on the tractor is currently being analyzed to develop easy-to-use and safe leverage that should be installed in the tractor cockpit. This will allow us to better analyze the feasibility of the proposed solutions and their testing during working activities, i.e., when the tractor is used in in-field operations. Another criticality is related to the fact that the implementation of these systems has been verified for one type of tractor only, while further dimensioning to adapt them to other tractor models is needed.

9. Conclusions

Nowadays, agricultural machinery safety still represents a criticality both considering occupational contexts (i.e., the use of work equipment in agricultural companies) and private activities (i.e., the use of agricultural machinery by so-called hobbyists). For this reason, the inclusion of ergonomics analyses to improve the safety and usability of tractors can be beneficial in both contexts, contributing to increasing the social aspects of the sustainability of agricultural systems.

More specifically, the current study is an attempt to reduce the research gap on the misuse of FROPSs, which has been highlighted by numerous authors, improving the results of previous studies through the development of two novel technical solutions to augment the safety level of farmers.

Hence, the results achieved can augment knowledge on FROPS assistance systems for agricultural tractors on the one hand, while on the other providing a methodical approach that can be used to include the ergonomics and human factors analysis in the development of technical solutions to upgrade this type of machinery in accordance with current safety requirements.

However, further research is being carried out to fully validate the systems and complete the feasibility analysis both for the analyzed tractor model and for other types.

Author Contributions: Conceptualization, D.G., L.V., D.P. and M.F.; methodology, D.G., L.V., D.P. and M.F.; software, D.G. and L.V.; validation, D.G., L.V., D.P. and M.F.; writing—review and editing, D.G., L.V., D.P. and M.F. All authors have read and agreed to the published version of the manuscript.

Funding: This research received no external funding.

Data Availability Statement: Not applicable.

Conflicts of Interest: The authors declare no conflict of interest.

References

1. Nawaz, W.; Linke, P.; Koç, M. Safety and sustainability nexus: A review and appraisal. *J. Clean. Prod.* **2019**, *216*, 74–87. [CrossRef]
2. Fargnoli, M.; Lombardi, M. NOSACQ-50 for safety climate assessment in agricultural activities: A case study in Central Italy. *Int. J. Environ. Res. Public Health* **2020**, *17*, 9177. [CrossRef] [PubMed]
3. Barneo-Alcántara, M.; Díaz-Pérez, M.; Gómez-Galán, M.; Carreño-Ortega, Á.; Callejón-Ferre, Á.J. Musculoskeletal disorders in agriculture: A review from Web of Science core collection. *Agronomy* **2021**, *11*, 2017. [CrossRef]
4. International Labour Office (ILO). Safety and Health at Work, A Vision for Sustainable Prevention. Available online: https://www.ilo.org/wcmsp5/groups/public/@ed_protect/@protrav/@safework/documents/publication/wcms_301214.pdf (accessed on 19 November 2022).
5. Facchinetti, D.; Santoro, S.; Galli, L.E.; Pessina, D. Agricultural tractor roll-over related fatalities in Italy: Results from a 12 years Analysis. *Sustainability* **2021**, *13*, 4536. [CrossRef]
6. Pochi, D.; Fornaciari, L.; Vassalini, G.; Grilli, R.; Fanigliulo, R. Levels of whole-body vibrations transmitted to the driver of a tractor equipped with self-levelling cab during soil primary tillage. *AgriEngineering* **2022**, *4*, 695–706. [CrossRef]
7. Vita, L.; Gattamelata, D.; Pessina, D. Retrofitting agricultural self-propelled machines with roll-over and tip-over protective structures. *Safety* **2021**, *7*, 46. [CrossRef]
8. Pinzke, S.; Svennefelt, C.A.; Lundqvist, P. Occupational injuries in Swedish agriculture–development and preventive actions. *J. Agric. Saf. Health* **2018**, *24*, 193–211. [CrossRef]
9. Myers, J.R.; Hendricks, K.J. Agricultural tractor overturn deaths: Assessment of trends and risk factors. *Am. J. Ind. Med.* **2010**, *53*, 662–672. [CrossRef]
10. Fargnoli, M.; Lombardi, M. Safety vision of agricultural tractors: An engineering perspective based on recent studies (2009–2019). *Safety* **2020**, *6*, 1. [CrossRef]
11. Sorensen, J.A.; Jenkins, P.; Bayes, B.; Clark, S.; May, J.J. Cost-effectiveness of a ROPS social marketing campaign. *J. Agric. Saf. Health* **2010**, *16*, 31–40. [CrossRef]
12. Hard, D.L.; McKenzie, E.A.; Cantis, D.; May, J.; Sorensen, J.; Bayes, B.; Madden, E.; Stone, B.; Maass, J. The NIOSH CROPS demonstration project: A study in New York and Virginia with an emphasis on youth. *J. Agric. Saf. Health* **2016**, *22*, 173–186. [CrossRef]
13. Cavallo, E.; Langle, T.; Bueno, D.; Tsukamoto, S.; Görücü, S.; Murphy, D. Rollover protective structure (ROPS) retrofitting on agricultural tractors: Goals and approaches in different countries. *J. Agromed.* **2014**, *19*, 208–209. [CrossRef]
14. European Agency for Safety and Health at Work (EU-OSHA). *Maintenance in Agriculture—A Safety and Health Guide*; Publications Office of the European Union: Luxembourg, 2011; Available online: https://data.europa.eu/doi/10.2802/54188 (accessed on 19 November 2022).
15. Gattamelata, D.; Puri, D.; Vita, L.; Fargnoli, M.; Lombardi, M. Retrofitting agricultural tractors with aftermarket weather cabins: Safety issues for manufacturers and users. In Proceedings of the International Conference on Industrial Engineering and Operations Management (IEOM), Rome, Italy, 2–5 August 2021; pp. 339–350. Available online: http://ieomsociety.org/proceedings/2021rome/306.pdf (accessed on 19 November 2022).
16. Mann, A.J.; Jepsen, S.D. Using expert panel data to guide youth agricultural safety and health training resources in the US. *Safety* **2017**, *3*, 4. [CrossRef]
17. Silleli, H.; Dayıoglu, M.A.; Gültekin, A.; Saranlı, G.; Yıldız, M.A.; Akay, E.; Ekmekçi, K. Anchor mechanism to increase the operator clearance zone on narrow-track wheeled agricultural tractors: Static and field upset test results. *Biosyst. Eng.* **2008**, *99*, 196–204. [CrossRef]
18. Ballesteros, T.; Arana, I.; Ezcurdia, A.P.; Alfaro, J.R. E2D-ROPS: Development and tests of an automatically deployable, in height and width, front-mounted ROPS for narrow-track tractors. *Biosyst. Eng.* **2013**, *116*, 1–14. [CrossRef]
19. Ballesteros, T.; Arana, J.I.; De Ezcurdia, A.P.; Alfaro, J.R. Development and validation of automatically deployable ROPS based on airbag inflator technology. *Biosyst. Eng.* **2015**, *130*, 92–105. [CrossRef]
20. Cremasco, M.M.; Caffaro, F.; Giustetto, A.; Vigoroso, L.; Paletto, G.; Cavallo, E. Tractor rollover protection: Is the incorrect use of foldable rollover protective structures due to human or to technical issues? *Hum. Factors J. Hum. Factors Ergon. Soc.* **2020**, *62*, 64–76. [CrossRef]

21. Vigoroso, L.; Caffaro, F.; Micheletti Cremasco, M.; Cavallo, E. Improving tractor safety: A comparison between the usability of a conventional and enhanced rear-mounted foldable ROPS (FROPS). *Int. J. Environ. Res. Public Health* **2022**, *19*, 10195. [CrossRef]
22. Molari, G.; Rondelli, V. On the definition of narrow-track wheeled agricultural tractors. *Biosyst. Eng.* **2004**, *88*, 75–80. [CrossRef]
23. Micheletti Cremasco, M.; Vigoroso, L.; Caffaro, F.; Paletto, G.; Cavallo, E. Considering human variability in the design of safe interaction with agricultural machinery: The case of foldable roll-over protective structure (FROPS) manual handling. *Agronomy* **2021**, *11*, 1303. [CrossRef]
24. Etherton, J.; McKenzie, J.E.A.; Lutz, T.; Cantis, D.; Kau, T.Y. An initial farmer evaluation of a NIOSH AutoROPS prototype. *Int. J. Ind. Ergon.* **2004**, *34*, 155–165. [CrossRef]
25. Alkhaledi, K.; Means, K.; McKenzie, J.E.; Smith, J. Reducing occupational fatalities by using NIOSH 3rd generation automat-ically deployable rollover protective structure. *Saf. Sci.* **2013**, *51*, 427–431. [CrossRef]
26. Pessina, D.; Facchinetti, D. A survey on fatal accidents for overturning of agricultural tractors in Italy. *Chem. Eng. Trans.* **2017**, *58*, 79–84. [CrossRef]
27. Gattamelata, D.; Vita, L.; Fargnoli, M. Machinery safety and ergonomics: A case study research to augment agricultural tracklaying tractors' safety and usability. *Int. J. Environ. Res. Public Health* **2021**, *18*, 8643. [CrossRef] [PubMed]
28. Fargnoli, M.; Vita, L.; Gattamelata, D.; Laurendi, V.; Tronci, M. A reverse engineering approach to enhance machinery design for safety. In Proceedings of the DESIGN 2012, the 12th International Design Conference, Dubrovnik, Croatia, 21–24 May 2012; Marjanovic, D., Storga, M., Pavkovic, N., Bojcetic, N., Eds.; International Design Conference: Dubrovnik, Croatia, 2012; pp. 627–636. ISBN 978-953-7738-17-4.
29. Wood, K.L.; Jensen, D.; Bezdek, J.; Otto, K.N. Reverse engineering and redesign: Courses to incrementally and systematically teach design. *J. Eng. Educ.* **2001**, *90*, 363–374. [CrossRef]
30. Organisation for the Economic Co-Operation and Development (OECD). Tractors standard Codes 4, 6, 7, 8. *Organisation for the Economic Co-Operation and Development Paris, France.* Available online: https://www.oecd.org/agriculture/tractors/codes/ (accessed on 19 November 2022).
31. Pessina, D.; Facchinetti, D.; Giordano, D.M. Narrow-track agricultural tractors: A survey on the load of the hand-operated foldable rollbar. *J. Agric. Saf. Health* **2016**, *22*, 275–284. [CrossRef]
32. Franceschetti, B.; Rondelli, V. Models to predict the force to operate front foldable rollover protective structures for narrow-track tractors. *Biosys. Eng.* **2019**, *185*, 126–134. [CrossRef]
33. Vigoroso, L.; Caffaro, F.; Cremasco, M.M.; Giustetto, A.; Paletto, G.; Cavallo, E. A bottom-up approach to tractor safety: Improving the handling of foldable roll-over protective structures (FROPS) through user-centred design. In *Innovative Bio-Systems Engineering for Sustainable Agriculture, Forestry and Food Production, Lecture Notes in Civil Engineering 67, Proceedings of the International Mid-Term Conference of the Italian Association of Agricultural Engineering, Matera, Italy, 12–13 September 2019*; Coppola, A., Di Renzo, G., Altieri, G., D'Antonio, P., Eds.; Springer Nature: Geneve, Switzerland, 2020; pp. 645–652. [CrossRef]
34. Vigoroso, L.; Caffaro, F.; Cavallo, E.; Cremasco, M.M. User-centred design to promote the effective use of rear-mounted foldable roll-over protective structures (FROPSs): Prototype evaluation among novice and expert farmers. *Span. J. Agric. Res.* **2021**, *19*, e0207. [CrossRef]
35. MSC Adams Software. Available online: https://hexagon.com/it/products/product-groups/computer-aided-engineering-software/adams (accessed on 19 November 2022).
36. Italian Workers' Compensation Authority (INAIL). L'installazione dei Dispositivi di Protezione in Caso di Ribaltamento nei Trattori Agricoli o Forestali. *Linee Guida: (Installation of Protection Devices in the Event of Overturning of Agricultural or Forestry Tractors. Guidelines).* Available online: https://www.inail.it/cs/internet/comunicazione/pubblicazioni/catalogo-generale/linstallazione-dei-dispositivi-di-protezione.html (accessed on 19 November 2022).
37. Fargnoli, M. Design for safety and human factors in industrial engineering: A review towards a unified framework. In Proceedings of the 11th Annual International Conference on Industrial Engineering and Operations Management, Singapore, 7–11 March 2021; pp. 7511–7522.
38. Urquhart, L.; Wodehouse, A.; Loudon, B.; Fingland, C. The application of generative algorithms in human-centered product development. *Appl. Sci.* **2022**, *12*, 3682. [CrossRef]
39. Kim, H.; Lee, K.; Räsänen, K. Agricultural injuries in Korea and errors in systems of safety. *Ann. Agric. Environ. Med.* **2016**, *23*, 432–436. [CrossRef]
40. Irwin, A.; Poots, J. Investigation of UK farmer go/no-go decisions in response to tractor-based risk scenarios. *J. Agromed.* **2018**, *23*, 154–165. [CrossRef]
41. Caffaro, F.; Cremasco, M.M.; Roccato, M.; Cavallo, E. Drivers of farmers' intention to adopt technological innovations in Italy: The role of information sources, perceived usefulness, and perceived ease of use. *J. R. Stud.* **2020**, *76*, 264–271. [CrossRef]
42. Fargnoli, M.; Lombardi, M.; Haber, N. A fuzzy-QFD approach for the enhancement of work equipment safety: A case study in the agriculture sector. *Int. J. Reliab. Saf.* **2018**, *12*, 306–326. [CrossRef]
43. Sadeghi, L.; Dantan, J.Y.; Mathieu, L.; Siadat, A.; Aghelinejad, M.M. A design approach for safety based on product-service systems and function–behavior–structure. *CIRP J. Manuf. Sci. Technol.* **2017**, *9*, 44–56. [CrossRef]
44. Keskin, S.G.; Keskin, M.; Soysal, Y. Assessing farm tractor incidents and awareness levels of operators for tractor safety issues in the Hatay province of Turkey. *J. Agric. Saf. Health* **2012**, *18*, 113–128. [CrossRef]

45. Kogler, R.; Quendler, E.; Boxberger, J. Occupational accidents with agricultural machinery in Austria. *J. Agromed.* **2016**, *21*, 61–70. [CrossRef]
46. Casazza, C.; Martelli, R.; Rondelli, V. Evaluation of a commercial tractor safety monitoring system using a reverse engineering procedure. *J. Agric. Saf. Health* **2016**, *22*, 215–225. [CrossRef]
47. Urbanic, R.J.; ElMaraghy, W. A design recovery framework for mechanical components. *J. Eng. Des.* **2009**, *20*, 195–215. [CrossRef]
48. Tinc, P.J.; Gadomski, A.; Sorensen, J.A.; Weinehall, L.; Jenkins, P.; Lindvall, K. Applying the consolidated framework for implementation research to agricultural safety and health: Barriers, facilitators, and evaluation opportunities. *Saf. Sci.* **2018**, *107*, 99–108. [CrossRef]

AgriEngineering

Article

Reduction in Atmospheric Particulate Matter by Green Hedges in a Wind Tunnel

Marcello Biocca *, **Daniele Pochi**, **Giancarlo Imperi** and **Pietro Gallo**

Consiglio per la Ricerca in Agricoltura e l'Analisi dell'Economia Agraria—CREA, Centro di Ricerca Ingegneria e Trasformazioni Agroalimentari, Via della Pascolare 16, 00015 Monterotondo, Italy; daniele.pochi@crea.gov.it (D.P.); giancarlo.imperi@crea.gov.it (G.I.); pietro.gallo@crea.gov.it (P.G.)

* Correspondence: marcello.biocca@crea.gov.it

Abstract: Urban vegetation plays a crucial role in reducing atmospheric particulate matter (PM), modifying microclimates, and improving air quality. This study investigates the impact of a laurel hedge (*Laurus nobilis* L.) on airborne PM, specifically total suspended particulate (TSP) and respirable particles (PM_4) generated by a Diesel tractor engine. Conducted in a wind tunnel of approximately 20 m, the research provides insights into dust deposition under near-real-world conditions, marking, to our knowledge, the first exploration in a wind tunnel of this scale. Potted laurel plants, standing around 2.5 m tall, were arranged to create barriers of three different densities, and air dust concentrations were detected at 1, 4, 9, and 14 m from the plants. The study aimed both to develop an experimental system and to assess the laurel hedge's ability to reduce atmospheric PM. Results show an overall reduction in air PM concentrations (up to 39%) due to the presence of the hedge. The highest value of dust reduction on respirable particles was caused by the thickest hedge (three rows of plants). However, the data exhibit varying correlations with hedge density. This study provides empirical findings regarding the interaction between dust and vegetation, offering insights for designing effective hedge combinations in terms of size and porosity to mitigate airborne particulate matter.

Keywords: urban forestry; air pollution; PM capture; particle deposition; porosity

check for updates

Citation: Biocca, M.; Pochi, D.; Imperi, G.; Gallo, P. Reduction in Atmospheric Particulate Matter by Green Hedges in a Wind Tunnel. *AgriEngineering* **2024**, *6*, 228–239. https://doi.org/10.3390/agriengineering6010014

Academic Editor: Mathew G. Pelletier

Received: 13 November 2023
Revised: 15 December 2023
Accepted: 17 January 2024
Published: 22 January 2024

1. Introduction

The impressive urban and landscape transformations of recent decades have corresponded to a dramatic growth of the urban population. Currently, more than half of the world's population (56.2%) lives in urban areas—progressively more in highly-populated cities [1]. In Italy, 69.5% of the population is urban [2]. Sustainable management of cities is a crucial point in stimulating the global ecological transition and mitigating climate change.

A major environmental health concern in urban areas is air pollution. The reduction in atmospheric pollutants and airborne particulate matter (PM) (or atmospheric dust) is one of the positive effects that urban vegetation provides to improve the quality of life. The term atmospheric dust refers to a mixture of solid and liquid particles suspended in the air which vary in size, composition, and origin. Some of the particles that make up atmospheric dust are emitted by various natural and anthropogenic sources ("primary particles"), while others derive from a series of chemical and physical reactions that occur in the atmosphere ("secondary particles"). There are also various removal mechanisms that the dust undergoes, including mechanisms that "remove" the dust from the atmospheric environment by causing it to fall back to the ground or towards the aquatic environment, as well as deposition mechanisms and active or passive obstacles for their interception [3].

The basis of dust classification is made up of the diameter of the particles (directly related to the possible interactions with the human body and, more specifically, with the respiratory and cardiovascular system) and their concentration. PM_{10} (with an average aerodynamic diameter of less than 10 μm) represents the dust capable of penetrating the upper respiratory tract, while $PM_{2.5}$ represents the dust able to penetrate the lower tract of the respiratory system (lung alveoli), causing serious harm to human health [4,5]. The health and quality of life of people living in cities are seriously threatened by air pollution. It is estimated that, in the year 2020, there were 52,300 premature deaths in Italy due to air exposure to $PM_{2.5}$ [6]. All over the world, the $PM_{2.5}$-related global premature mortality is estimated to be around 4.3 and 4.4 million [7].

Several measures can be set up to reduce particulate matter emissions and mitigate their effects, including the presence of vegetation, which plays a key role in contrasting the air pollution by means of different mechanisms. Numerous studies have shown that the plants offer an effective method of air purification, especially for non-point source particulate pollution, because of their wide distribution [8,9].

At least three major types of green infrastructure contribute to this effect: trees and hedges (which make up the so-called vertical greenery), surfaces (such as turf), and green roofs and walls. In particular, hedges represent an important and widespread form of greenery in the city. Although they vary greatly in terms of the species used, vegetative habitus, size, and density of foliage, hedges have a common trait: they are generally located in proximity to the main sources of city pollution, such as vehicular traffic. Therefore, hedges can be considered the first line of defense against smog emissions resulting from vehicular traffic. Hedges are also characterized by having a rather homogeneous morphology, with continuous vegetation from the base to the apex of the crown and a shape that can be traced back to a parallelepiped or a cuboid.

The main reason for the presence of hedges in urban environments lies in their ability to function as visual barriers. For example, hedges are placed in traffic reservations to reduce the nuisance of glare caused by vehicles passing in the opposite lane, around gardens and houses to increase privacy, and around sports facilities to help athletes focus. Their great aesthetic function is undeniable, including their presence in historic gardens and parks, where they may represent a distinctive feature, as seen in formal or Italian gardens. Together with their aesthetic functions, hedges contribute to the urban green endowment and the supply of ecosystem services typical of urban greenery, such as mitigation of the heat island effect [10], reduction in noise pollution [11,12], increase in biodiversity [13], and improvement of the microclimate [14]. Authors have even studied the role of urban hedges in blast mitigation [15]. Hedges are also attractive as they contribute to the resilience and adaptability of urban forests due the fact that they increase structural diversity in terms of age, spatial profile, and species distribution. Structural (and, thus, functional) diversity can be achieved quickly by planting appropriate shrubs in addition to slower-growing trees [16].

It is worth noting that hedges, unlike tree cover, can reduce air pollution even in places characterized by the presence of tall buildings around relatively narrow roads (also known as "road canyons"), where air ventilation and dilution of pollutants may be insufficient [17]. In these situations, the cover of tree crowns may sometimes cause a local accumulation of dust and pollutants, while the presence of hedges does not have this effect. Results have indicated that different kinds of greenbelts can improve footpath air quality to a certain degree (7–15%). Interestingly, the vegetation structure of shrubs and small trees (about 2–4.5 m in height) with small crown diameters shows the highest PM_{10} removal efficiency along major or heavy traffic roads [18,19]. In open road locations, most studies report a reduction in the concentrations of various airborne pollutants due to hedgerows in the range of 15–60% [20]. Other authors have assessed the pollutant removal capacity in relation to botanical species [21,22]. Moreover, the habitus of the species (evergreen or broadleaves) influences the total yearly capacity of dust interception, which is stronger in evergreen than in broad-leaved plants.

Our research focused on the ability of hedges to remove particulate matter or atmospheric dust. Many authors have studied removal mechanisms and assessed the capacity of green barriers to reduce particulate matter using different techniques. Experimental methods include field experiments, chamber experiments, modelling and computer simulations, remote sensing, and wind tunnels. Wind tunnels are experimental setups that simulate the flow of air in a controlled environment over a specific area. Generally, air containing particulate matter is introduced into the tunnel and passed over a section of vegetation. The air is then measured for the concentration of PM both before and after it has passed through the vegetation to quantify the amount of PM that has been removed by the vegetation. However, wind tunnels are typically small-scale and may not accurately reflect real-world conditions [23–25]. Despite these limitations, wind tunnels can still provide valuable information about the reduction in airborne particulate matter by vegetation and can be a useful complement to field and computer simulation studies. Overall, a combination of experimental approaches, including wind tunnel studies, can help to provide a comprehensive understanding of the role of vegetation in reducing airborne particulate matter [9,26,27].

When planning and designing hedges for urban areas, landscape architects, agronomists, and urban planners must deal with multiple factors to optimize the positioning of plants, to determine the right botanical species, and to choose the size and density of the vegetal barrier [28]. Elements related to optimization of these parameters are valued in order to minimize planting costs and maximize the ecological benefits of vegetations [29].

The aims of our work were:

- To set up an experimental system that allows for an assessment of the ability of hedges to reduce atmospheric particulate matter;
- To evaluate the relationship between hedge density and dust reduction ability;
- To quantify the air concentration of PM at different distances from green barriers;
- To provide useful elements to design hedges for urban areas.

The experimental setting consisted of a wind tunnel ca. 20 m long and of green barriers made up of mature living laurel plants (*Laurus nobilis* L.). This resulted in an original arrangement which has never been employed before. To our knowledge, this is the first study to be carried out in a wind tunnel of such a size.

2. Materials and Methods

In recent years, a work area was created in our institute to evaluate the dispersion of abrasion dust from seeders that simulated the sowing of dressed seeds of maize. Numerous tests were carried out in this facility, and the results were validated through appropriate open field tests [30,31]. The same experimental setup was then utilized in the present study to evaluate the reduction in and the drift of airborne PM incited by a green barrier. The wind tunnel was arranged at the workshop's porch of our institute to obtain a site that was protected by external influences and large enough to contain the hedge and measure dust drift. The test area measured 6.7 m wide, more than 20 m long, and 5.0 m high.

The external open side of the porch was closed by means of tarpaulin sheets settled as curtains (Figure 1). Consequently, the site was closed on each side and the end of the "gallery" was left open to permit the air to flow. At the test site, artificial wind conditions were produced by means of an electric axial fan (0.8 m diameter) operating at 1700 rpm. Preliminary tests were carried out with the aim of verifying the repeatability and the constancy of wind conditions (speed and direction) at the test site by means of a portable anemometer (Schiltknecht, Schaffhausen, Switzerland, Micro MiniAir 4). The wind speed was measured at a height of 0.7 m from the ground at 21 points along the test area.

The diagram in Figure 1 shows the arrangement of the relevant instruments, the source of emissions, the fan, and the green barrier.

The green barrier was made by placing the plants in containers (0.25 m in diameter) side by side to form one, two, or three rows of plants, in order to obtain hedges of different thicknesses and densities (Figure 2). The hedges were placed at 6.0 m from the fan.

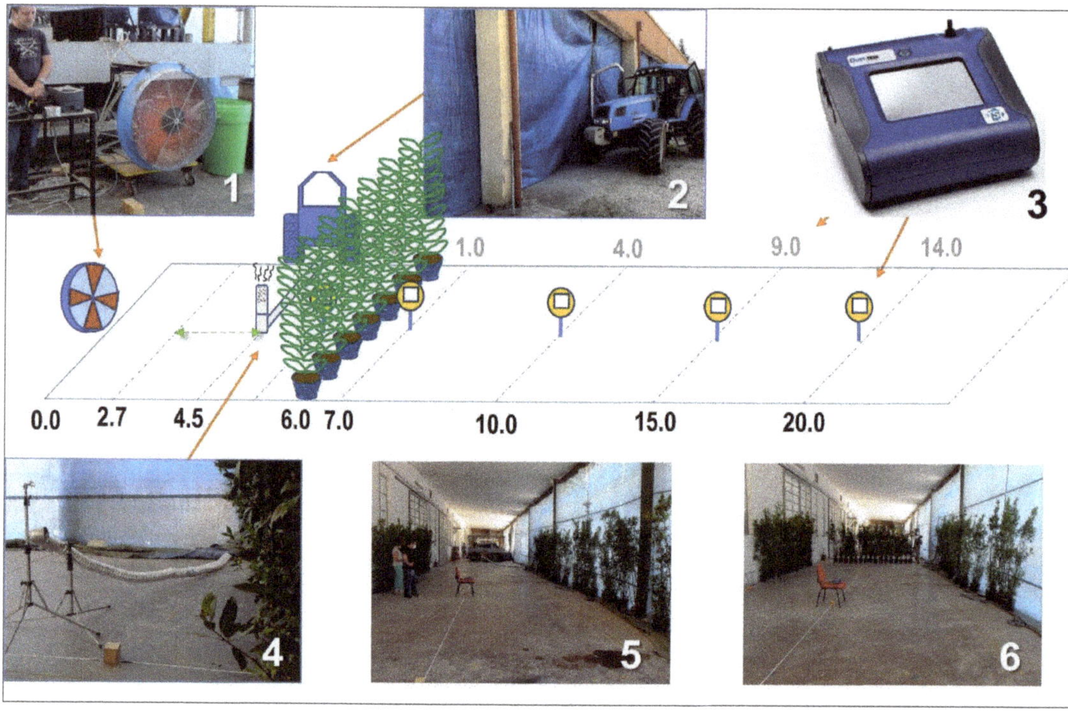

Figure 1. Diagram of the experimental area. Numbers are distances (in meters) of the sampling points from the fan (black character) and from the hedge (grey character). (**1**) The fan to generate the air flow; (**2**) the tractor placed outside the wind tunnel with the pipe to collect the exhaust gas; (**3**) the "DustTrak" portable photometer (TSI, Shoreview, MN, USA); (**4**) the point where the exhaust gas of the tractor is released in the test area; (**5**) and (**6**) views of the wind tunnel with and without the hedge.

Number of pot rows	Thickness (m)	Height (m)	Width (m)	Porosity (%)	Pots pattern
1	~ 0.40			14.3	
2	~ 0.85	2.4	5.5	4.6	
3	~ 1.20			1.4	

Figure 2. Main characteristics of the hedges.

The optical porosity of the hedge was the parameter chosen to describe its density. This was determined through the analysis of digital images processed with ImageJ software, version 1.53e (Figure 3) [32].

The concentrations of two fractions of PM were then measured: the total suspended particulate (TSP, that is the fraction containing particles with particle diameters <50–100 μm) and the respirable particles (PM_4). Measurement of the PM concentration was carried out with a TSI/Tecora "DustTrak" portable photometer capable of detecting, by mounting a suitable cutting head, the concentration of particulate matter (expressed in mg m^{-3}) of the two different size classes. Each acquisition consisted of 30 measures, with a frequency equal to 1 Hz. During sampling, the recorded concentration values were constantly monitored to ensure that data acquisition did not present excessive variability. The sampler operated at a flow rate of 1.7 L min^{-1} for PM_4 and 3.0 L min^{-1} for TSP. Before each experiment, the sampling area was carefully cleaned to minimize interferences arising from pre-existing dust.

Figure 3. Steps of image analysis to estimate the porosity of the hedge. (**1**) Image of the vegetation; (**2**) black and white 8-bit image transformation; (**3**) segmentation and analysis (percentage of the two areas).

Sampling was carried out at 1.0, 4.0, 9.0, and 14.0 m from the hedge, correspondent to 7.0, 10.0, 15.0, and 20.0 m from the fan, at a height of 0.7 m from the ground. Additional samplings of PM were carried in proximity to the gas bypass exit (1.0 m from the outlet) to measure the PM concentration before the hedge.

The source of the particulate matter was the exhaust fumes of a 145 HP Landini Legend tractor (Landini, Reggio Emilia, Italy) operating at 1500 rpm. The tractor was placed outside the wind tunnel, and the exhaust gases were intercepted via a pipe which released them at a height of 0.7 m from the ground, close to the hedge, at 4.5 m from the fan used for the formation of the air flow.

The TSP reduction was also tested in an additional test, with the pipe emitting the exhaust gases placed at 2.7 m from the fan, farther from the vegetation. As mentioned, samplings were repeated at the same points in the absence of the plants and with hedges formed by one, two, and three rows of plants.

In summary, the experimental design entailed PM_4 and TSP sampling with 4 hedge configurations at 4 distances from the barrier, and with the fan placed at 2 distances from the hedge. Since each sampling was equal to 30 measures, we obtained a data set of 1820 data points.

Removal efficiency, R, was calculated using the following equation:

$$R[\%] = \frac{PM\ concentration\ without\ plants\ \left[\mathrm{mg\ m^{-3}}\right]\ -\ PM\ concentration\ with\ plants\ \left[\mathrm{mg\ m^{-3}}\right]}{PM\ concentration\ without\ plants\ \left[\mathrm{mg\ m^{-3}}\right]} \times 100 \quad (1)$$

Statistical analysis to determine significant differences between the PM removal capacities of hedges with different densities were performed with a bifactorial ANOVA (analysis of variance) parametric test, with the distance of sampling points and hedge density as factors. Since the variable was not normally distributed (Shapiro–Wilk test, p-value < 0.01), it was normalized according to the results of a Box–Cox test procedure, which showed that the best technique of data normalization was the logarithmic transformation. The analyses were performed using R software, version 1.53e [33].

3. Results and Discussion

The air velocity inside the wind tunnel sampling area was measured downwind of the hedge, where the average wind speed (measured at a height of 0.7 m from the ground) along the test area was 1.2 m s^{-1} in the absence of the hedge and 0.5 m s^{-1} with the hedge. No substantial differences in average wind velocity were recorded with the hedge formed by one, two, or three rows of plants. The obtained pattern of the air flow is shown in Figure 4, where it is compared the air pattern velocity in the sampling area without plants and with a hedge formed by three rows of pots. The plants induced an average reduction in wind speed, which decreased to zero after approximately 9 m from the hedge. However, around 15 m from the hedge, a slight increase in air velocity was observable (Figure 4B).

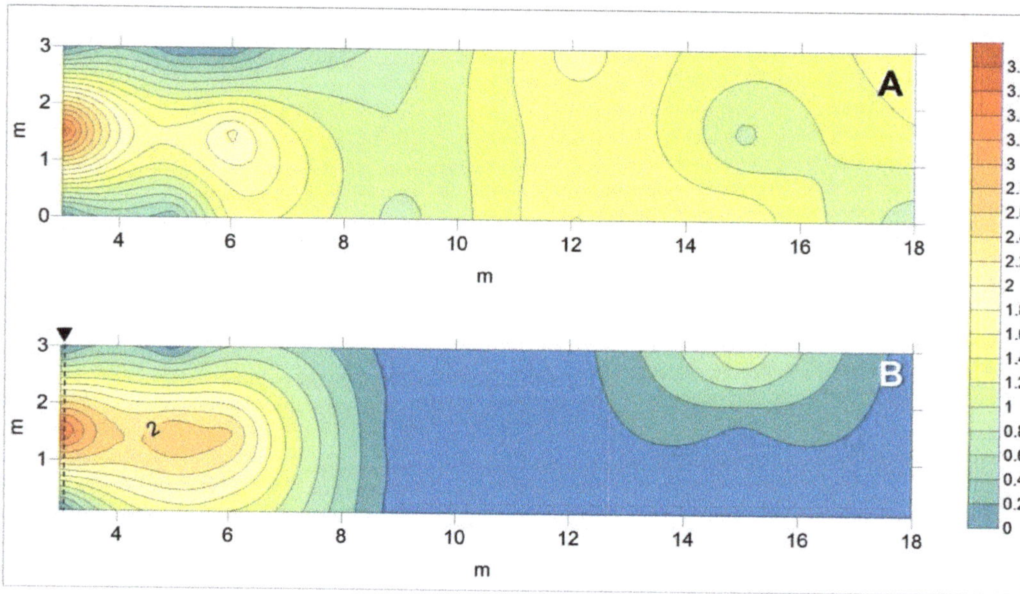

Figure 4. Contour maps of air wind patterns in the empty wind tunnel, without plants (**A**) and with a hedge of three rows of plants (**B**). The marked line in (**B**) indicates the positions of plants. Values are expressed in m s^{-1}.

The initial concentrations of particles were measured in proximity to the exhaust gas outlet coming from the tractor. The values, reported in Table 1, showed a high concentration of fine particles in the composition of the aerosol.

Table 1. Average air concentration values (mg m^{-3} $\pm$ standard error) of PM in proximity to the gas exit (the PM_4 was not tested at 2.7 m).

Fraction Size	Distance of the Fan from the Gas Exit					
	4.5 m			2.7		
TSP	0.83	±	0.06	0.64	±	0.02
PM_4	4.73	±	0.08	-		-

Regarding the PM reduction detected in the test area, the results are reported in Tables 2–4.

Table 2. Reduction in TSP air concentration due to the hedge with respect to the concentration without plants in the test area. Exhaust gas exit was placed at 4.5 m from the fan.

	Distance of Sampling Points from the Fan (Distance of Sampling Points from the Plants)				
Row of Plants	**7 (1) m**	**10 (4) m**	**15 (9) m**	**20 (14) m**	**Average**
1	−9%	−30%	27%	39%	7%
2	13%	−3%	78%	6%	24%
3	49%	61%	3%	41%	38%

Table 3. Reduction in PM_4 air concentration due to the hedge with respect to the concentration without plants in the test area. Exhaust gas exit was placed at 4.5 m from the fan.

	Distance of Sampling Points from the Fan (Distance of Sampling Points from the Plants)				
Row of Plants	**7 (1) m**	**10 (4) m**	**15 (9) m**	**20 (14) m**	**Average**
1	21%	53%	52%	32%	39%
2	−16%	82%	60%	22%	37%
3	54%	15%	44%	52%	41%

Table 4. Reduction in TSP air concentration due to the hedge with respect to the concentration without plants in the test area. Exhaust gas exit was placed at 2.7 m from the fan.

	Distance of Sampling Points from the Fan (Distance of Sampling Points from the Plants)				
Row of Plants	**7 (1) m**	**10 (4) m**	**15 (9) m**	**20 (14) m**	**Average**
1	−65%	48%	46%	52%	20%
2	0%	26%	23%	39%	22%
3	−31%	74%	32%	−84%	−2%

Descriptive statistics (median, 25th and 75th percentiles, minimum and maximum, outliers) of the dust concentrations in the sampling area for each distance from the hedge are reported in Figure 5. Air concentration values in mg m^{-3} are shown in Table S1 (Supplementary Material). As expected, the PM concentration values in the air correlated with the sampling distance, with partial or no reduction just behind the plants (1 m from the hedge) and maximum reductions in the rest of the sampling area. The ANOVA results, reported in Table 5, showed significant differences according to both of the considered factors, i.e., distance and hedge density.

Table 5. Reduction in TSP air concentration due to the hedge with respect to the concentration without plants in the test area. Exhaust gas exit was placed at 2.7 m from the fan.

Test	Factors	Degrees of Freedom	Probality (p-Values)	Significance (1)
PM_4	Hedge rows	3	<0.001	***
	Distance	3	<0.001	***
	Hedge rows × Distance	9	<0.001	***
TSP with fan at 4.5 m	Hedge rows	3	<0.001	***
	Distance	3	<0.001	***
	Hedge rows × Distance	9	<0.001	***
TSP with fan at 2.7 m	Hedge rows	3	<0.001	***
	Distance	3	<0.001	***
	Hedge rows × Distance	9	<0.001	***

(1)—Significance codes of p-values: 0 '***'; 0.001

The maximum overall reductions (41% and 38%) were recorded for PM_4 in the case of the thickest hedge (three rows of plants).

A single row of plants can cause a reduction in TSP air concentration only around 9 m from the plant, while in the immediate downwind zone, closer to the hedge, the values are higher. The reduction in PM_4, on the other hand, appears to have been more consistent in the various cases.

When the exhaust gas outlet was moved away from the hedge and closer to the fan, the overall TSP reduction pattern significantly changed. In this case, at a distance of 14 m from the hedge, a peak concentration of PM was recorded with the thickest hedge, while the hedges with one or two rows of plants showed a certain abatement effectiveness. As mentioned previously, this set (exhaust gas exit placed at 2.7 m from the fan) was not tested for the finest fraction.

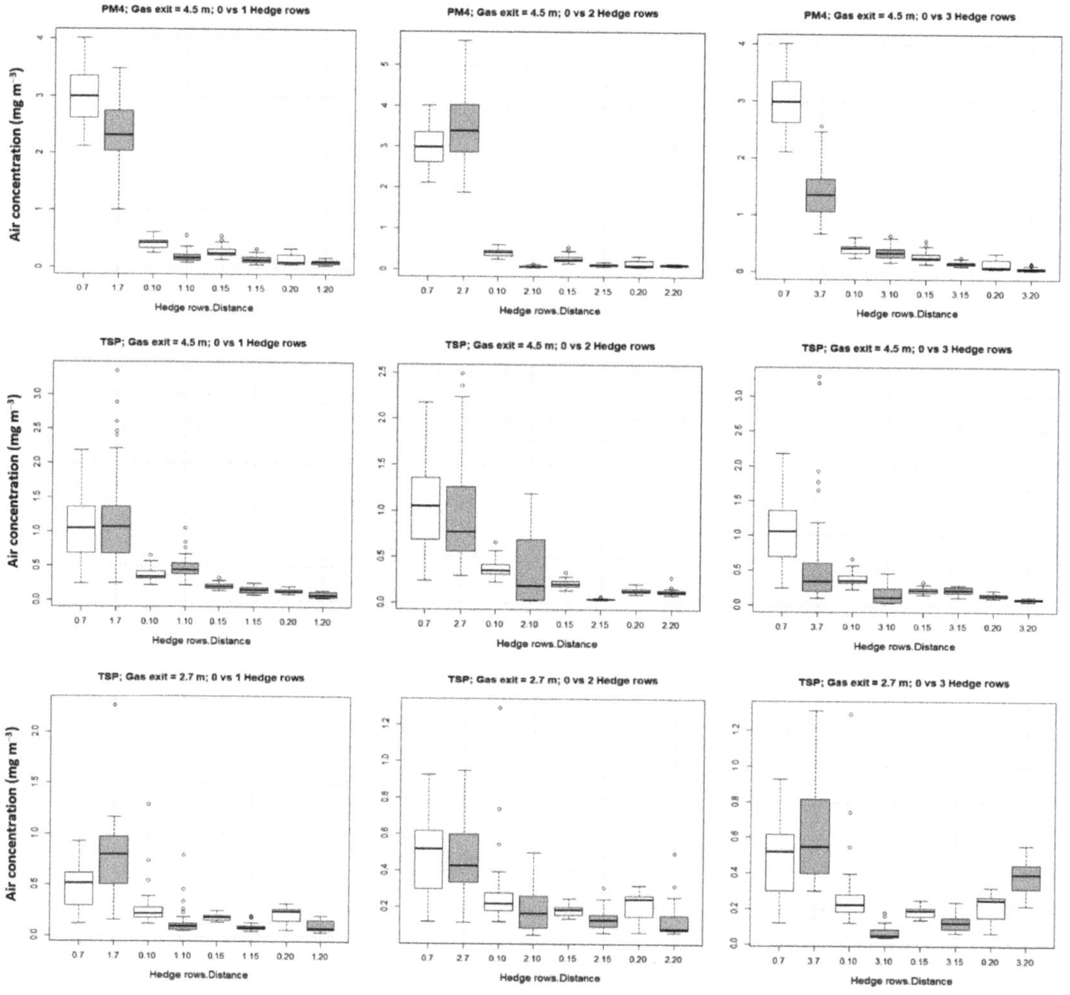

Figure 5. Comparison between the values of PM_4 and TSP concentrations in the test area, with and without hedges, by the distance. The codes on the horizontal axis show the number of hedge rows and the distances in meters (line within box: median; box limits: 25th and 75th percentiles; whisker ends: minimum and maximum; empty circles: outliers).

The denser screen (three rows), even if, in general, it is more effective in filtering the dust in the various experiments, appeared to be of little effectiveness in the case in which the source of the exhaust gas was moved away from the green screen and closer to the fan (exhaust gas exit placed at 2.7 m from the fan). In this case, it is very likely that the air flow was not able to penetrate the hedge, and thus passed over the hedge. In fact, the wind speed pattern in the sampling area (Figure 4) shows an increase around 15 m. This hypothesis is also supported by the observation that the worst average performance occurred away from the hedge, at the sampling point located at 20 m.

The observed values of overall dust concentration reduction were strictly correlated with the edge porosity (showed in Table 2) in only one test, specifically in the case of TSP assessment with the fan placed at 4.5 m from the exhaust gas exit (Figure 6). In the case of PM_4, a relation was not observed because the reduction values were similar with each hedge.

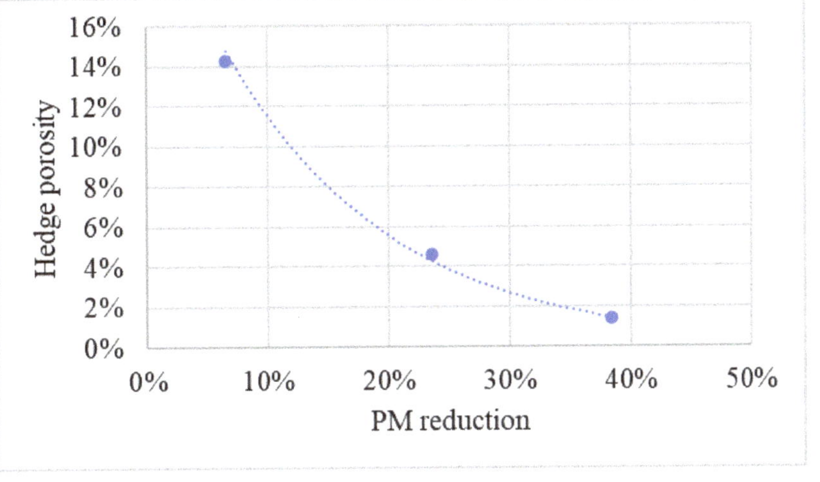

Figure 6. Reduction in TSP by the variation in hedge porosity (statistics of the curve: y = 0.2383 $e^{-7.289x}$; R^2 = 0.9987).

This implies that the total deposition is determined by a trade-off, as stated by Raupach and collaborators [34]: "the windbreak must be dense enough to absorb particles efficiently but sparse enough to allow some particles to flow through and be trapped". Other studies reached similar conclusions thanks to CFD modeling, which highlighted the behavior of particle passing vegetation of different densities in relation to their size [35].

4. Conclusions

Wind tunnels are a valuable tool for studying the reduction in airborne particulate matter by vegetation, and many authors have used these facilities. However, wind tunnel experiments may not fully replicate real-world conditions, and the results obtained in a wind tunnel environment may not be translated directly to the field. In particular, small wind tunnels allow for better control of the different factors involved, but the findings are less generalizable. In this study, the size of the wind tunnel allowed for tests of large structures, like those of a real hedge, allowing us to sample the PM up to 15–20 m from the emission source.

We measured the reduction in airborne particulate matter (PM) caused by laurel plants and compared the air concentrations of PM_4 and TSP (total suspended particulate) with and without the presence of hedges. On average, reductions in airborne particulate matter concentration were observed with denser screens, but results were also affected by other factors.

We confirmed that the phenomenon was influenced by numerous and complex relations among air movements, vegetation (density, morphology, species, size, etc.) PM concentration, and particle size distribution. The obtained data were affected by multiple involved factors, and for this reason, they often appear very variable. It is likely that, in addition, resuspended dust episodes could affect PM air concentrations independently from the hedge thickness.

A distinctive aspect of this study Is the measurement of dust deposition at varying distances from the green barrier, offering insights into the potential area affected by dust deposition. Such information could be useful in designing and planning hedge placement.

A limit of this study concerns the use of a single plant species, the laurel, a species that is, in any case, widely used to grow hedges in Mediterranean cities.

The comparison of TSP concentrations between the two sets with different distances of the gas exit from the hedge (2.7 and 4.5) showed that denser hedges can act as barriers, which limit the concentration just behind the hedge but have no effect in terms of reducing the overall dust concentration. In this case, dust can jump the hedge and redeposit after some meters. Therefore, the development of a new green structure must consider its density (and dimensions) as a key factor.

In conclusion, it is necessary to underscore the broader significance of urban vegetation in contributing to global sustainability. By enhancing air quality and mitigating the adverse effects of airborne particulate matter, urban vegetation plays a crucial role in fostering a sustainable and resilient environment.

Supplementary Materials: The following supporting information can be downloaded at: https://www.mdpi.com/article/10.3390/agriengineering6010014/s1, Table S1: Air concentration values in mg m^{-3}.

Author Contributions: Conceptualization, M.B. and P.G.; methodology, M.B, G.I. and D.P.; validation, M.B., G.I. and P.G.; investigation, M.B. and P.G.; resources, M.B.; data curation, M.B.; writing—original draft preparation, M.B.; writing—review and editing, M.B.; visualization, M.B.; supervision, M.B. and P.G.; funding acquisition, M.B. All authors have read and agreed to the published version of the manuscript.

Funding: This research was funded by the Italian national project URBANFOR3, funded by Lazio Innova (CUP: C82I16000000005).

Data Availability Statement: The data presented in this study are available on request from the corresponding author.

Acknowledgments: The authors are grateful to Beatrice Bassotti of CREA for her administrative support.

Conflicts of Interest: The authors declare no conflicts of interest.

References

1. UN-Habitat. *World Cities Report 2020: The Value of Sustainable Urbanization*; UN-Habitat: Nairobi, Kenya, 2020; 418p.
2. Worldometer. 2020, Italy Population. Available online: https://www.worldometers.info/world-population/italy-population/ (accessed on 10 December 2023).
3. Litschke, T.; Kuttler, W. On the reduction of urban particle concentration by vegetation—A review. *Meteorol. Z.* **2008**, *17*, 229–240. [CrossRef]
4. Donaldson, K.; Seaton, A. A short history of the toxicology of inhaled particles. *Part. Fibre Toxicol.* **2012**, *9*, 13. [CrossRef] [PubMed]
5. Li, N.; Hao, M.; Phalen, R.F.; Hinds, W.C.; Nel, A.E. Particulate air pollutants and asthma: A paradigm for the role of oxidative stress in PM-induced adverse health effects. *Clin. Immunol.* **2003**, *109*, 250–265. [CrossRef] [PubMed]

6. EEA, European Environment Agency. Health Impacts of Air Pollution in Europe, 2022. Web Report. Available online: www.eea.europa.eu/publications/air-quality-in-europe-2022/health-impacts-of-air-pollution (accessed on 27 February 2023).
7. Im, U.; Bauer, S.E.; Frohn, L.M.; Geels, C.; Tsigaridis, K.; Brandt, J. Present-day and future $PM_{2.5}$ and O_3-related global and regional premature mortality in the EVAv6.0 health impact assessment model. *Environ. Res.* **2023**, *216 Pt 4*, 114702. [CrossRef] [PubMed]
8. Beckett, K.P.; Freer-Smith, P.H.; Taylor, G. Urban woodlands: Their role in reducing the effects of particulate pollution. *Environ. Pollut.* **1998**, *99*, 347–360. [CrossRef]
9. Janhäll, S. Review on urban vegetation and particle air pollution—Deposition and dispersion. *Atmos. Environ.* **2015**, *105*, 130–137. [CrossRef]
10. Viecco, M.; Jorquera, H.; Sharma, A.; Bustamante, W.; Fernando, H.J.S.; Vera, S. Green roofs and green walls layouts for improved urban air quality by mitigating particulate matter. *Build. Environ.* **2021**, *204*, 108120. [CrossRef]
11. Van Renterghem, T.; Attenborough, K.; Maennel, M.; Defrance, J.; Horoshenkov, K.; Kang, J.; Bashir, I.; Taherzadeh, S.; Altreuther, B.; Khan, A.; et al. Measured light vehicle noise reduction by hedges. *Appl. Acoust.* **2014**, *78*, 19–27. [CrossRef]
12. Biocca, M.; Gallo, P.; Di Loreto, G.; Imperi, G.; Pochi, D.; Fornaciari, L. Noise attenuation provided by hedges. *J. Agric. Eng.* **2019**, *889*, 113–119. [CrossRef]
13. Alvey, A.A. Promoting and preserving biodiversity in the urban forest. Urban Forest. *Urban For. Urban Green.* **2006**, *5*, 195–201. [CrossRef]
14. Escobedo, F.J.; Kroeger, T.; Wagner, J.E. Urban forests and pollution mitigation: Analyzing ecosystem services and disservices. *Environ. Pollut.* **2011**, *159*, 2078–2087. [CrossRef]
15. Gajewski, T.; Peksa, P.; Studziński, R.; Malendowski, M.; Sumelka, W.; Sielicki, W.P. Application verification of blast mitigation through the use of thuja hedges. *Int. J. Prot. Struct.* **2022**, *13*, 363–378. [CrossRef]
16. FAO. *Guidelines on Urban and Peri-Urban Forestry*; Salbitano, F., Borelli, S., Conigliaro, M., Chen, Y., Eds.; FAO Forestry Paper No. 178.; Rome, Food and Agriculture Organization of the United Nations: Rome, Italy, 2016.
17. Wania, A.; Bruse, M.; Blond, N.; Weber, C. Analysing the influence of different street vegetation on traffic-induced particle dispersion using microscale simulations. *J. Environ. Manag.* **2012**, *94*, 91–101. [CrossRef]
18. Shen, J.; Gao, Z.; Ding, W.; Yu, Y. An investigation on the effect of street morphology to ambient air quality using six real-world cases. *Atmos. Environ.* **2017**, *164*, 85–101. [CrossRef]
19. Chen, X.; Pei, T.; Zhou, Z.; Teng, M.; He, L.; Luo, M.; Liu, X. Efficiency differences of roadside greenbelts with three configurations in removing coarse particles (PM_{10}): A street scale investigation in Wuhan, China. *Urban For. Urban Green.* **2015**, *14*, 354–360. [CrossRef]
20. Abhijith, K.V.; Kumar, P.; Gallagher, J.; McNabola, A.; Baldauf, R.; Pilla, F.; Broderick, B.; Di Sabatino, S.; Pulvirenti, B. Air pollution abatement performances of green infrastructure in open road and built-up street canyon environments—A review. *Atmos. Environ.* **2017**, *162*, 71–86. [CrossRef]
21. Mori, J.; Sæbø, A.; Hanslin, H.M.; Teani, A.; Ferrini, F.; Fini, A.; Burchi, G. Deposition of traffic-related air pollutants on leaves of six evergreen shrub species during a Mediterranean summer season. *Urban For. Urban Green.* **2015**, *14*, 264–273. [CrossRef]
22. Vigevani, I.; Corsini, D.; Mori, J.; Pasquinelli, A.; Gibin, M.; Comin, S.; Szwałko, P.; Cagnolati, E.; Ferrini, F.; Fini, A. Particulate Pollution Capture by Seventeen Woody Species Growing in Parks or along Roads in Two European Cities. *Sustainability* **2022**, *14*, 1113. [CrossRef]
23. Guo, L.; Zhao, B.; Zhao, D.; Li, J.; Tong, J.; Chang, Z.; Liu, X. Evaluation of conifer and broad-leaved barriers in intercepting particulate matters in a wind tunnel. *J. Air Waste Manag. Assoc.* **2020**, *70*, 1314–1323. [CrossRef]
24. Chiam, Z.; Song, X.P.; Lai, H.R.; Tan, H.T.W. Particulate matter mitigation via plants: Understanding complex relationships with leaf traits. *Sci. Total Environ.* **2019**, *688*, 398–408. [CrossRef]
25. Lin, M.-Y.; Khlystov, A. Investigation of Ultrafine Particle Deposition to Vegetation Branches in a Wind Tunnel. *Aerosol Sci. Technol.* **2012**, *46*, 465–472. [CrossRef]
26. Xie, C.; Kan, L.; Guo, J.; Jin, S.; Li, Z.; Chen, D.; Li, X.; Che, S. A dynamic processes study of PM retention by trees under different wind conditions. *Environ. Pollut.* **2018**, *233*, 315–322. [CrossRef] [PubMed]
27. Ysebaert, T.; Koch, K.; Samson, R.; Denys, S. Green walls for mitigating urban particulate matter pollution—A review. *Urban For. Urban Green.* **2021**, *59*, 127014. [CrossRef]
28. Varshney, C.K.; Mitra, I. Importance of hedges in improving urban air quality. *Landsc. Urban Plan.* **1993**, *25*, 75–83. [CrossRef]
29. Ghafari, S.; Kaviani, B.; Sedaghathoor, S.; Allahyari, M.S. Ecological potentials of trees, shrubs and hedge species for urban green spaces by multi criteria decision making. *Urban For. Urban Green.* **2020**, *55*, 126824. [CrossRef]
30. Biocca, M.; Conte, E.; Pulcini, P.; Marinelli, E.; Pochi, D. Sowing simulation tests of a pneumatic drill equipped with systems aimed at reducing the emission of abrasion dust from maize dressed seed. *J. Environ. Sci. Health Part B* **2011**, *46*, 438–448.
31. Biocca, M.; Fanigliulo, R.; Gallo, P.; Pulcini, P.; Pochi, D. The assessment of dust drift from pneumatic drills using static tests and in-field validation. *Crop Prot.* **2015**, *71*, 109–115. [CrossRef]
32. Schneider, C.A.; Rasband, W.S.; Eliceiri, K.W. NIH Image to ImageJ: 25 years of image analysis. *Nat. Methods* **2012**, *9*, 671–675. [CrossRef]

33. R Core Team. *R: A Language and Environment for Statistical Computing*; R Foundation for Statistical Computing: Vienna, Austria, 2022. Available online: https://www.R-project.org (accessed on 12 October 2023).
34. Raupach, M.R.; Woods, N.; Dorr, G.; Leys, J.F.; Cleugh, H.A. The entrapment of particles by windbreaks. *Atmos. Environ.* **2001**, *35*, 3373–3383. [CrossRef]
35. Tiwary, A.; Morvan, H.P.; Colls, J.J. Modelling the size-dependent collection efficiency of hedgerows for ambient aerosols. *J. Aerosol Sci.* **2006**, *37*, 990–1015. [CrossRef]

AgriEngineering

MDPI

Article

Levels of Whole-Body Vibrations Transmitted to the Driver of a Tractor Equipped with Self-Levelling Cab during Soil Primary Tillage

Daniele Pochi, Laura Fornaciari, Gennaro Vassalini, Renato Grilli and Roberto Fanigliulo *

Consiglio per la Ricerca in Agricoltura e l'Analisi dell'Economia Agraria (CREA), Centro di Ricerca Ingegneria e Trasformazioni Agroalimentari (Research Centre for Engineering and Agro-Food Processing), Via della Pascolare 16, Monterotondo, 00015 Rome, Italy; daniele.pochi@crea.gov.it (D.P.); laura.fornaciari@crea.gov.it (L.F.); gennaro.vassalini@crea.gov.it (G.V.); renato.grilli@crea.gov.it (R.G.)

* Correspondence: roberto.fanigliulo@crea.gov.it; Tel.: +39-06-90675232

Abstract: Agricultural tractor drivers' health preservation and comfort represent important aspects of the evolution of agricultural machinery and led to the development of devices aimed at improving working conditions, such as soundproof cab and driver seat suspension, nowadays commonly adopted in tractors. The vibrations are one of the factors mostly affecting health and comfort conditions, resulting from the characteristics and interaction of specific tractor's parts (tyres, axles, chassis, cab). Trying to improve their products, manufacturers developed a cab prototype equipped with an automatic self-levelling system, whose goal is to maintain the driver's vertebral column in a correct position during heavy agricultural operations such as primary soil tillage. A tractor with a such a prototype was tested to assess its effectiveness in maintaining the cab horizontal and any effects on the transmitted levels of whole-body vibration, during soil primary tillage carried out by means of a mouldboard plough and a subsoiling plough, both in plain and hilly surfaces. The results showed that the device worked well at a slope lower than the operating limits of the system, keeping the cabin horizontal through progressive adjustments. A slight reduction of the level of vibration was observed with a self-levelling system working during the tillage tests in the plain, compared to the traditional condition.

Keywords: health preservation; whole-body vibrations; daily exposure time; mouldboard plough; subsoiling plough

Citation: Pochi, D.; Fornaciari, L.; Vassalini, G.; Grilli, R.; Fanigliulo, R. Levels of Whole-Body Vibrations Transmitted to the Driver of a Tractor Equipped with Self-Levelling Cab during Soil Primary Tillage. *AgriEngineering* **2022**, *4*, 695–706. https://doi.org/10.3390/agriengineering4030044

Academic Editor: Francesco Marinello

Received: 25 May 2022
Accepted: 28 July 2022
Published: 1 August 2022

Publisher's Note: MDPI stays neutral with regard to jurisdictional claims in published maps and institutional affiliations.

1. Introduction

In recent years, beside the operative performances, the attention of the manufacturers of agricultural machinery has been increasingly focused on the aspects of comfort and health preservation of the operator, leading to the introduction of devices and instruments capable of significantly improving the working conditions. For example, modern agricultural tractor cabs are normally equipped with air conditioning, soundproofing systems [1] and driver seats with suspensions effective against the vibrations typical of agricultural work, responsible for temporary discomfort conditions. The vibrations transmitted to the driver's whole-body (WBV) are one of the factors most affecting health and comfort conditions. They are the result of the characteristics of elements such as cabin, tyres, chassis [2–4], axles [5,6], and seat suspensions [7–9] which differently interact depending on the external conditions (soil unevenness, slope, type of agricultural operation, speed, etc.).

In this regard, the manufacturers of tractors are trying to improve the health safeguard and comfort conditions in their products, through the development and introduction of new devices. Among them we find the cabin equipped with an automatic self-levelling system, the goal of which is to change the cockpit setting and maintain the driver in the correct position during agricultural operations that are more demanding on his spine. Driving tractors, in fact, induces a postural overload, accentuated by any transverse slopes, due to the frequent rotations of the lumbar spine to perform operations in a sitting position [10].

Many researchers demonstrated that musculo-skeletal disorders (MSDs) in the driver workplace are caused by mechanical vibrations, and that there is a relationship between WBV exposure and MSDs, especially for low back pain (LBP), associated with increased risk of injury [11–14]. Few studies on WBV exposure in actual tractor-driving conditions are available [15,16]. Such risks become dangerous when the intensity of the vibrations is high, includes strong shocks or jolts, and occurs frequently with prolonged and repetitive exposure [17]. According to the Directive 2002/44/EC, tractor drivers' exposure to WBV should not exceed a daily action value of 0.5 m s^{-2} (exposure action value, EAV) [18]. The cited Directive and the European Parliament established the minimum protection requirements for the workers exposed to the risks of vibration in the workplace.

When the WBV exceeds the EAV, actions to reduce the risks from vibration must be taken. On the other hand, the level of WBV is affected by the type of operation, the relating implement coupled to the tractor and the surface characteristics. For example, after harvesting, the presence of clods and deep cracks makes the soil remarkably hard and difficult to till and sow, particularly on slopes [19–21]. Primary tillage foresees the inversion of the soil layers, by means of reversible mouldboard plough, or a vertical shatter without mixing the tilled layers, carried out by implements such as a subsoiling plough, which aims to restore soil structure and to mitigate soil compaction [22,23]. The subsoiling is a principal substitutive of the ploughing and, differently from the ploughing, it favours the formation of the soil structure reducing superficial cloddiness [24], often making harrowing unnecessary.

This study concerned a series of tests on a medium-power tractor featuring a prototype of automated self-levelling cab with hydraulic control, which should allow the operator to maintain the correct vertical posture even during heavy tillage, such as deep ploughing in-furrow and subsoiling, on both plain and hilly surfaces, helping to reduce the effects of prolonged exposure to vibration on the spine in incorrect posture. The study was focused on the aspect of the driver's health to verify the effectiveness of the elastic systems in damping vibrations that are harmful to the human body. Therefore, the levels of vibration transmitted from the seat to the driver were measured according to the standard ISO 2631-1:1997 [25] in the frequency band 0.5–80 Hz considered more dangerous for the human body in a sitting position. The standard also specifies the location and direction of the measurements, the equipment to be used, the duration of the measurements and the frequency weighting, the measurement assessment methods and the evaluation of weighted root-mean-square (RSM) acceleration. The tests also aimed at determining the best efficiency conditions in maintaining the cab horizontal, the self-levelling cab operating limits, the speed of intervention, any effects on the levels of vibration. The ploughing and subsoiling operations were found to exhibit vibration exposure at low frequencies in the vicinity of natural frequencies of the human body and may consequently affect a driver's health and comfort [26].

2. Materials and Methods

2.1. The Tractor Used in the Tests

The tests were conducted on a tractor of medium–high power (147 kW) equipped with an automated self-levelling cab and with a high level of comfort for the driver. To provide the updated characteristic curves of the tractor's engine, aimed at quantifying the power available, the tractor was connected at a dynamometric brake (Borghi and Saveri, Bologna, Italy) [27]. Considering the subject of the tests, beyond the driver seat with pneumatic suspension (in which a rubber cylinder containing pressurized air works as a spring), the tractor had, as its most peculiar characteristic, an original hydraulic system operating the self-levelling of the cab, powered by the oil of the hydraulic system of the tractor (Figure 1).

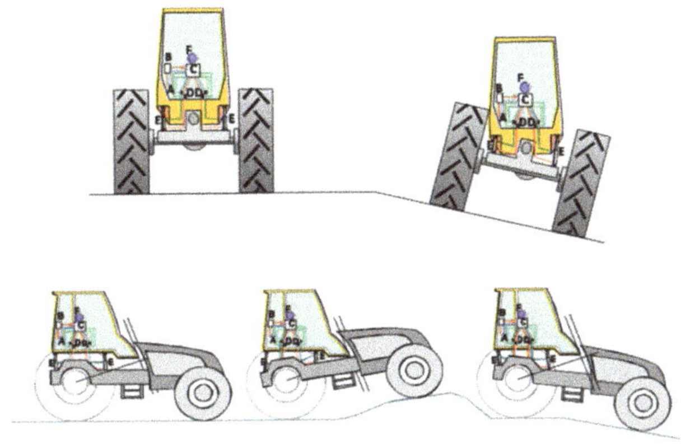

Figure 1. Sketch of the self-levelling cab according to the transversal and longitudinal slopes: (A) Oil flow from the reservoir and back; (B) hydraulic pump; (C) hydraulic distributor; (D) solenoid valves; (E) double effect cylinders; (F) gyroscopic sensor and computer.

The self-levelling apparatus is based on the presence of four hydraulic cylinders at the four corners of the cab floor, to form a square of 0.90 m side. A gyroscope detects changes in the slope of the ground according to the longitudinal and transverse directions and controls the action of the cylinders in such a way that the cab always remains horizontal. The maximum range of the cylinders, measured with the tractor stationary, is 0.23 m, corresponding to a maximum gradient of 25.5% and a maximum angle of 14.3° with respect to the two axes of the horizontal plane. The four cylinders are independent of one another and represent the only connection between cab and frame. The engine also has no point of contact with the interior. A silent block is mounted on each cylinder, with the function of reducing the vibrations, especially the high frequency ones (Figure 2). Therefore, the cabin, during the work, always remained parallel to a reference plane that forms a zero angle with the horizontal.

Figure 2. Particulars of a hydraulic cylinder and of the silent block on its top.

2.2. Instruments

The instrumental chain was composed of:

- Two six-channel signal conditioners Brüel & and Kjær;
- Eight-channel digital recorder;
- Signal acquisition and processing system Brüel & Kjær 5/1-ch. Input/Output Controller Module 0 Hz to 25.6 kHz frequency range (Figure 3a). The used sampling frequency was 160 Hz—suitable for analysing the level of vibration on tractors during field operations;

- Tri-axial accelerometer adapted for driver seat Brüel & Kjær, type 4322 (Figure 3b) with relative calibrator, type 4294;
- Two tri-axial accelerometers Brüel & Kjær, type 4321, positioned on the tractor chassis and on the cab floor (Figure 3c).

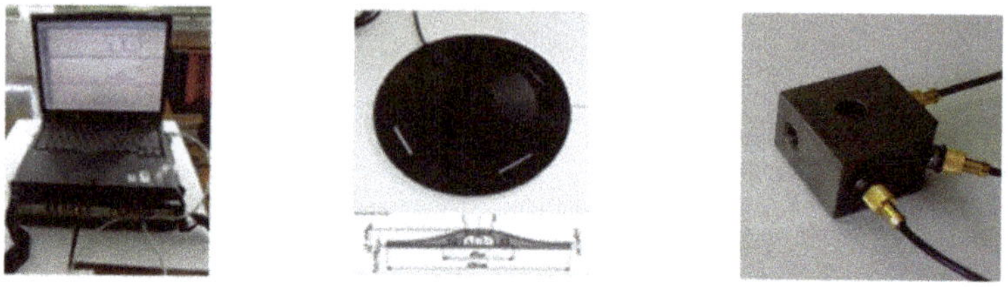

(a) (b) (c)

Figure 3. (**a**) Signal acquisition and processing system for data; (**b**) Three-axial accelerometer for driver seat; (**c**) Three-axial accelerometer placed on the cab floor.

2.3. Measured Parameters, Data Processing and Reference Parameters

The tests aimed to verify the presence of any effects due to the cab self-levelling system on the level of vibration transmitted to the whole-body of the tractor driver. The basic parameter to measure is the acceleration, a, expressed in m s^{-2}. As the effects of the vibrations depend on the frequency of the accelerations, these must be weighted by means of suitable filters according to the standard ISO 2631-1:1997, in order to quantify the WBV exposure in the reference frequency band 0.5–80 Hz, which is more dangerous for the human body in a sitting position. To observe how the vibrations are transmitted from the soil to the driver's seat through the various tractor elements, a tri-axial accelerometer was fixed on the driver's seat and the remaining two were placed, respectively, on the cab floor and on the tractor main frame, taking care that all three lay on the same vertical line. Vibration must be detected on three axes, defined by a coordinate system referring to the human body and originating at the point of contact between the subject and the vibrating surface. The *x*-axis passes through the back and chest, the *y*-axis through the shoulders, the *z*-axis through the feet and head (buttocks and head in the case of a seated person). As regards vehicle drivers, the x, y and z axes coincide with the longitudinal, transverse and vertical axes of the vehicle respectively. By adopting the risk assessment criterion for the health of seated subjects, according to the standard, the driver's seat was taken as the measurement plane and the triaxial seat accelerometer was positioned there, as shown in Figure 4.

Figure 4. The driver seat equipped with tri-axial accelerometer and system of coordinates for the orientation of the accelerometer for seated position.

As regards the health, the evaluation criterion defined by the ISO 2631-1 standard refers to the various pathologies produced by vibrations and in particular to alterations of the spine. The criterion considers subjects regularly exposed to vibrations and concerns individuals in a seated position. The effects on health depend on the dose of vibrations absorbed, therefore the same effects correspond to the same dose.

Beyond the evaluation of vibrations with reference to their effects on health, the cited standard also states how to evaluate them in terms of effects on comfort, referring in this case to the means of transport where it is influenced by multiple factors. The interference between vibrations and certain activities (such as reading, writing, and drinking) may sometimes represent a cause of discomfort and involves the aspect of perception.

The comfort can be assessed through specific measurement of the accelerations on the three axes, on the seat cushion, on the seat back and under the feet. The ISO 2631-1 standard reports a vibration assessment scale formulated by passengers on public transport. The values refer to the overall sum of the vibrations.

Since these criteria can hardly be extended to our test conditions, it was decided to focus the analysis of the potential effects of the self-levelling cabin on the reduction of vibration levels harmful to human health, postponing the in-depth analysis on the effects on comfort to a later stage.

Since the effects of vibrations vary depending on the frequency of the acceleration, this must be weighted in frequency. The sensitivity of the human body to vibrations is at a maximum within a certain frequency range and gradually decreases, moving away from its lower and upper limits. Therefore, the weighting works by letting the signal generated by the accelerometer (analysed in frequency) pass unaltered in the range of maximum sensitivity and in attenuating it to a more or less progressive extent externally. In our case, the W_d filter was adopted for the x and y axes, while the W_k filter was applied for the z axis. This processing provided, for each axis, the frequency weighted acceleration, a_w:

$$a_w = \left[\frac{1}{T}\int_0^T a_w^2(t)dt\right]^{1/2}, \tag{1}$$

where: a_w (t) is the measured value of the acceleration; T is the acquisition time interval (s). The components of the acceleration along the three axes are simultaneously measured. The resultant vector of the acceleration, a_v, is provided by the relation:

$$a_v = \left(k_x^2\, a_{wx}^2 + k_y^2 a_{wy}^2 + k_z^2 a_{wz}^2\right)^{1/2}, \tag{2}$$

where a_{wx}, a_{wy}, a_{wz} are weighted RMS accelerations along the x, y and z axes; k_x, k_y, k_z are indices the values of which were determined depending on the effects of the relative components of the acceleration on the health: for k_x e k_y a value of 1.4 is applied in the case of sitting positions, as they are equal to 1 for the upright position; k_z is equal to 1 in both positions.

The exposure to the vibrations can be quantified by normalizing the value of the acceleration a_v, measured during the daily exposure time (T_e), referring to an 8-h time interval, according to the principle of "equal energy", providing the normalized acceleration, A(8) according to the formula:

$$A(8) = a_w max\sqrt{\frac{T_e}{8}}. \tag{3}$$

However, as to the whole-body vibrations, the calculation of A(8) aimed at the health risk assessment is normally made only by considering the higher axial a_w component. The determination of A(8) and the calculation of the maximum daily exposure times were performed using an Excel datasheet.

2.4. Soil Tillage Tests

The tractor was used in soil tillage tests, choosing operations which, involving the inclination of conventional cabs such as in-furrow ploughing or working on slopes, emphasize the interventions of the self-levelling system. Therefore, the tractor was firstly employed in the execution of in-furrow ploughing at a depth of 0.35 m with a three mounted furrow plough, on both plain and hilly surfaces. Then, on the same surfaces of the ploughing tests, the tractor was used to operate a mounted subsoiling plough at a depth of 0.20 m (Figure 5), which represented a further severe test condition [28,29].

(a) (b)

Figure 5. The tractor equipped with the self-levelling cab during the tests with the plough (**a**) and the subsoiling plough (**b**).

As regards the hill trials, a very demanding trajectory was chosen, with slopes (longitudinal and transversal) varying up to a maximum of over 45%. The operations were carried out transversely with reference to the dominant hillside slope, to highlight the operational capabilities of the self-levelling system.

As for the gear box ratios used in the tests, they were chosen to maximize, in each condition, the working speed. The sampling time intervals had the following mean durations: ploughing in plain: 94.2 s; Subsoiling in plain: 68.0 s; Ploughing on slope: 80.9 s; Subsoiling on slope: 80.7 s. The operations were executed with the self-levelling system disconnected (conventional mode, "OFF mode") and connected (system working, "ON mode") with the aim of comparing the results of the measurements of the levels of vibrations. Three replications were made for each test condition.

3. Results and Discussion

The results of the measurements of the levels of vibration during the operations described above are shown in Table 1 and in Figures 6 and 7.

In Table 1 it can be observed that the peaks of instant axial accelerations always occurred in the low frequency range, with small differences among chassis, cab floor and seat. It can be noticed that in the X and Y axes frequencies from 1 Hz up to 1.6 Hz were involved, without evident differences between OFF/ON, ploughing/subsoiling, in-plain/hillside. Higher peak frequencies were observed in the Z axis, with values frequently ranging from 2.5 Hz up to 6 Hz. Moreover, the ploughing peak frequencies were higher than those of subsoiling and the hillside tillage peak frequencies were higher than in-plain tillage frequencies.

As regards the trend of the weighted r.m.s. axial accelerations, the component a_{wz} at the driver seat is always lower than a_{wx} and a_{wy} because of the damping action of the seat suspension on the Z axis.

Table 1. Tillage tests in plain and hillside. a_{wx}, a_{wy}, a_{wz}: weighted r.m.s. axial accelerations; Dom.: dominant component (in bold character), among a_{wx}, a_{wy}, a_{wz}, multiplied by 1.4; Res.: resultant vector of the acceleration, a_v; T_e: time of exposure as a function of "Dom." and "Res; a_{peak}: maximum observed instant acceleration; F_{peak}: frequency in correspondence of a_{peak}; ST: safety time (A(8) = 0.5 m s^{-2}); *LT*: limit time (A(8) = 1 m s^{-2}). The differences between ON and OFF modes are also reported for the accelerations (at chassis, cab floor and seat) and for ST and LT (only at driver seat).

Tillage	Mode	Sensor's Position	X Axis			Y Axis			Z Axis			Dom.	$T_{e\,(Dom.)}$		Res.	$T_{e\,(Res.)}$	
			a_{wx}	a_{peak}	F_{peak}	a_{wy}	a_{peak}	F_{peak}	a_{wz}	a_{peak}	F_{peak}	$1.4 \cdot a_{wmax}$	ST	LT	a_v	ST	LT
			(m s^{-2})		(Hz)	(m s^{-2})		(Hz)	(m s^{-2})		(Hz)	(m s^{-2})	(h:min)	(h:min)	(m s^{-2})	(h:min)	(h:min)
Ploughing in plain	**Off**	**Seat**	**0.38**	**0.22**	1.2	0.36	0.2	1.2	0.28	0.14	2.5	0.53	6:31	28:16	0.78	2:60	12.60
		Cab floor	0.3	0.16	1.2	0.29	0.17	1.2	0.29	0.13	2.5	0.42	-	-	0.65	-	-
		Chassis	0.28	0.15	1.2	0.23	0.14	1.2	0.28	0.12	2.5	0.39	-	-	0.58	-	-
	On	Seat	**0.37**	0.2	1.4	0.33	0.17	1.0	0.27	0.15	2.5	0.52	6:52	29:49	0.74	3:19	14:25
		Cab floor	0.26	0.13	1.4	0.24	0.14	1.0	0.3	0.13	2.5	0.30	-	-	0.58	-	-
		Chassis	0.4	0.26	1.1	0.17	0.1	1.0	0.28	0.12	2.5	0.56	-	-	0.67	-	-
	Diff. On-Off	Seat	−0.01	-	-	−0.03	-	-	−0.01	-	-	−0.01	0:21	1:33	−0.04	0:20	1:25
		Cab floor	−0.04	-	-	−0.05	-	-	0.01	-	-	−0.12	-	-	−0.07	-	-
		Chassis	0.12	-	-	−0.06	-	-	0	-	-	0.17	-	-	0.09	-	-
Subsoiling in plain	Off	Seat	**0.37**	0.22	1.2	0.35	0.28	1.7	0.19	0.1	1.0	0.52	6:52	29:49	0.74	3:23	14:42
		Cab floor	0.5	0.28	1.2	0.62	0.41	1.0	0.49	0.2	2.2	0.87	-	-	1.22	-	-
		Chassis	0.41	0.24	1.2	0.57	0.34	1.7	0.32	0.16	1.0	0.80	-	-	1.03	-	-
	On	Seat	**0.39**	0.25	1.3	0.23	0.17	1.6	0.15	0.08	1.0	0.55	6:11	26:50	0.65	4:21	18:51
		Cab floor	0.46	0.26	1.0	0.56	0.39	1.0	0.43	0.17	2.6	0.78	-	-	1.10	-	-
		Chassis	0.35	0.19	1.0	0.49	0.28	1.6	0.28	0.13	1.0	0.69	-	-	0.89	-	-
	Diff. On-Off	Seat	0.02	-	-	−0.12	-	-	−0.04	-	-	0.03	−0:41	−2:59	−0.09	0:58	4:10
		Cab floor	−0.04	-	-	−0.06	-	-	−0.06	-	-	−0.08	-	-	−0.12	-	-
		Chassis	−0.06	-	-	−0.08	-	-	−0.04	-	-	−0.11	-	-	−0.15	-	-
Hillside ploughing	Off	Seat	**0.34**	0.19	1.0	0.30	0.14	1.1	0.28	0.12	2.5	0.48	8:08	35:18	0.69	3:50	16:37
		Cab floor	0.24	0.13	1.0	0.24	0.14	1.1	0.27	0.11	5.0	0.27	-	-	0.55	-	-
		Chassis	0.21	0.11	1.0	0.21	0.14	1.0	0.27	0.11	5.0	0.27	-	-	0.50	-	-
	On	Seat	0.33	0.18	1.1	**0.35**	0.2	1.0	0.29	0.12	4.3	0.49	7:41	33:19	0.73	3:26	14:53
		Cab floor	0.25	0.13	1.2	0.31	0.2	1.0	0.34	0.14	6.3	0.34	-	-	0.65	-	-
		Chassis	0.21	0.1	1.2	0.22	0.16	1.0	0.3	0.13	5.0	0.30	-	-	0.52	-	-
	Diff. On-Off	Seat	−0.01	-	-	0.05	-	-	0.01	-	-	0.01	−0:27	−1:59	0.04	−0:24	−1:44
		Cab floor	0.01	-	-	0.07	-	-	0.07	-	-	0.07	-	-	0.11	-	-
		Chassis	0	-	-	0.01	-	-	0.03	-	-	0.03	-	-	0.03	-	-
Hillside subsoiling	Off	Seat	**0.34**	0.21	1.1	0.3	0.18	1.0	0.22	0.09	2.9	0.48	8:08	35:18	0.67	4:05	17:43
		Cab floor	0.22	0.14	1.0	0.24	0.16	1.0	0.19	0.08	5.0	0.34	-	-	0.49	-	-
		Chassis	0.18	0.11	1.0	0.18	0.13	1.0	0.2	0.08	5.0	0.20	-	-	0.41	-	-
	On	Seat	**0.36**	0.23	1.1	0.33	0.22	1.0	0.23	0.1	2.3	0.50	7:15	31:30	0.72	3:33	15:22
		Cab floor	0.22	0.15	1.1	0.26	0.19	1.0	0.2	0.08	4.0	0.36	-	-	0.52	-	-
		Chassis	0.18	0.12	1.1	0.18	0.13	1.0	0.2	0.08	5.0	0.20	-	-	0.41	-	-
	Diff. On-Off	Seat	0.02	-	-	0.03	-	-	0.01	-	-	0.03	−0:53	−3:49	0.05	−0:32	−2:21
		Cab floor	0	-	-	0.02	-	-	0.01	-	-	0.03	-	-	0.02	-	-
		Chassis	0	-	-	0	-	-	0	-	-	0.00	-	-	0.00	-	-

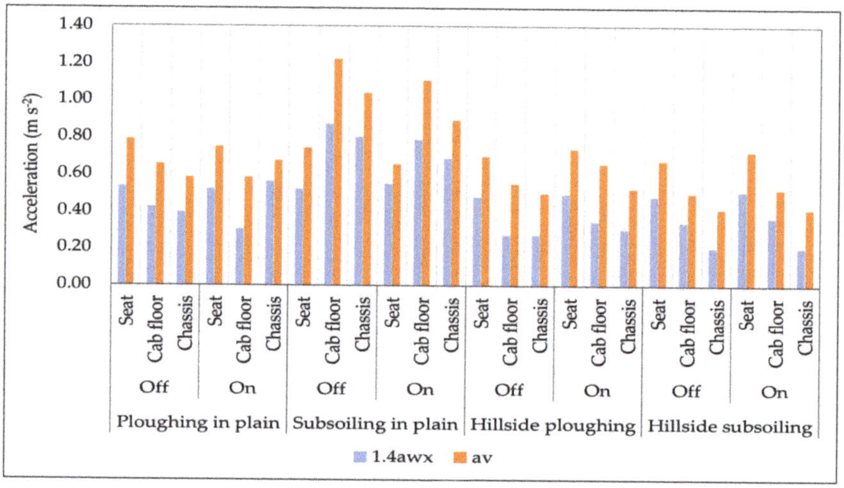

Figure 6. Average values of the vector acceleration, av, in the three points of application of the accelerometers in the different soil tillage.

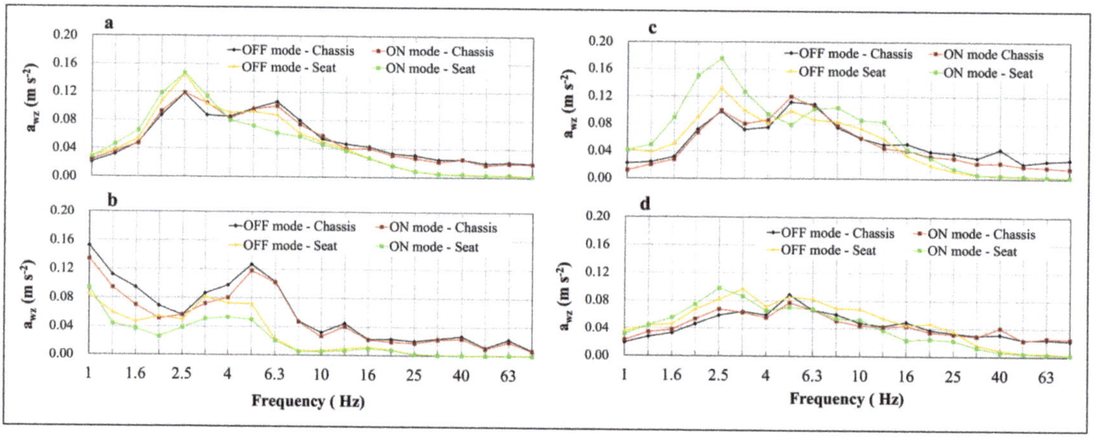

Figure 7. Frequency analysis of the acceleration along the Z axis during the four tillage tests, at the driver seat and at the chassis of the tractor. (**a**) ploughing in plain; (**b**) subsoiling in plain; (**c**) hillside ploughing; (**d**) hillside subsoiling.

Considering the ploughing in-plain, the levels of vibration at the driver seat are always lower in "ON mode" than in "OFF mode". In "ON mode", despite at frame level the components a_w resulted in being higher on all axes; the corresponding a_w values at the cab floor and at the seat resulted in being lower than in "OFF mode".

As to the hillside ploughing, despite similar a_w at the chassis in both modes, as a_{wx} decreased in ON mode, the combined effect of in-furrow ploughing, high soil cloddiness and cross slope caused an increase in a_{wy} and a_{wz} at cab floor and seat. This is particularly evident on the Y axis and is probably caused by the interventions of the self-levelling system to compensate to restore the cab horizontality, which could result in frequently not being adequate in terms of speed [30–32].

The vibration levels measured during the subsoiling in-plain show values of the components a_w at the chassis in ON mode far lower than in OFF mode, which makes the

a_w comparison between the two modes at cab floor and at seat level difficult. However, a_{wx} was higher in ON mode than in OFF mode, while a_{wy} and a_{wz} were higher in OFF mode.

During the hillside subsoiling, despite similar acceleration values observed at the chassis in both modes, the solicitations at the seat in ON mode increased on all axes, determining the worsening of the working conditions.

Table 1 shows, in bold characters, the components of the accelerations which were dominant in each test condition. The component on the X axis was dominant in all tests except in hillside ploughing, where a_{wy} prevailed. Probably, the variations in traction force during the tillage determined prevalent solicitations along the travel direction (the X axis), but in hillside ploughing in ON mode, they were overtaken by the transversal ones (Y axis) deriving from the interaction of transversal slope, in-furrow ploughing, great clods, and cabin movements operated by the self-levelling system. In fact, during this test, the slope sometimes exceeded the operational limit of the system (25.5% or 14.3°, see point 2.1) that was frequently disconnected and reconnected in an attempt to maintain the horizontality.

It can also be noticed, however, that in many cases the aforesaid dominants are not clearly higher than the other components whose values are often very close to those of the former. In particular, when a_{wx} is dominant, a_{wy} is the component with the closest value. Conversely, if a_{wy} prevails, a_{wx} follows it closely. This is important because of the higher relative weight of such components in WBV risk assessment according to the ISO 2631-1:1997. The values of a_{wz} are always more distant, probably due to the effective action exerted by the seat suspension on the Z axis. However, with respect to the dominants, they do not appear to be negligible as well.

For these reasons, in addition to the safety time (ST) and limit time (LT) commonly calculated, according to the above standard, basing on the dominant components, Table 1 also shows the ST and LT values based on the vector sum, a_v, as stated by the same standard when the vibration in two or more axes is comparable [33], in order to assess how the three components affect the exposure time. The A(8) reference values of 0.5 m s^{-2} and 1 m s^{-2} were adopted respectively as safety value and limit value.

In Figure 6, the dominant axial component (a_{wx} or a_{wy}) multiplied by the coefficient 1.4 is compared to the values of the resultant acceleration, a_v, calculated according to relation (2). It can be noticed that the latter is always significantly higher than the corresponding values of the former and, apart from these differences, that their trends are not always coherent in the different tests. For instance, in "subsoiling in plain", passing from OFF mode to ON mode, a_{wx} increases, while a_v decreases. This confirms that, with reference to the dominant, when non-dominant axial components are relatively high, they play an important role in determining the general level of vibration. Excluding them from the computation of the exposure time could lead to underestimation in WBV risk assessment and/or to different conclusions in the evaluation of an equipment such as the subject of this study. Said considerations clearly reflect on the values of ST and LT, which will result longer or shorter (Table 1) if they are calculated, respectively, on $1.4 \cdot a_{wmax}$ or, more prudentially, on a_v.

The diagrams of Figure 7 show the frequency analyses of the accelerations on the Z axis measured at the driver seat and at the chassis (which reflects the soil unevenness) during the four tillage tests.

The curves in OFF and ON mode at the chassis have similar trends in each diagram, as some differences can be observed at seat level and can be attributed to the action of the self-levelling system combined with the seat suspension. Another general indication is that the frequency of the peak acceleration at the seat is lower than the frequency of the peak acceleration at the chassis. In both ploughing tests, the peaks at the seat are higher than at the chassis [34]. Then, in particular:

- Ploughing in-plain (Figure 7a): the curves at the driver seat in OFF and ON modes are very similar with their peak at 2.5 Hz, with values higher than those at the chassis. Small differences can be noticed for frequencies lower than 4.0 Hz, where the accelerations for "ON mode" are slightly greater, and in the interval 4.0–12.5 Hz, where the values of "ON mode" are lower;

- Hillside ploughing (Figure 7c): the differences between OFF and ON have widened: "ON mode" clearly shows higher acceleration than OFF mode in the interval 1.0–4.0 Hz and 6.3–16.0 Hz. Additionally in this case, the peaks occurred at 2.5 Hz;
- Subsoiling in-plain (Figure 7b): even if the curves of the Z-acceleration at the chassis have similar shapes (with peaks at 5.0 Hz), below 6.3 Hz they have lower values in ON mode than in OFF. This is probably due to differences in soil unevenness and is reflected by the curves at the seat where, in the interval 1.0–6.3 Hz, the ON mode acceleration is much less than in "OFF mode". The shapes of the curves of the acceleration at the seat are different from those at the chassis, with peaks at 1.0 Hz in both theses;
- Hillside subsoiling (Figure 7d): in this case, the curves of the acceleration at the chassis have similar shapes (with peaks at 5.0 Hz), with small differences in the interval 1.0–6.3 Hz where the values of the "ON mode" are slightly higher than in A. At seat level, in the interval 1.0–5.0 Hz, the accelerations are higher than at the chassis and the "ON mode" shows worse behaviour (with peak at 2.5 Hz) than the "OFF mode" (with peak at 3.15 Hz).

4. Conclusions

The results of first test on a system for the self-levelling of an agricultural tractor cab indicate that it could contribute to improve the health preservation and comfort of the driver and, ultimately, to enhancing working conditions. The system has the function of maintaining the spine in a vertical position regardless of the slope of the ground, limiting damages resulting from stresses caused by uncomfortable posture (e.g., curvature of the lumbar region). The test highlighted both some operational situations in which the system improved the driver working conditions, and other situations that have been found to be critical for it and require further studies.

As to the mechanical performance, the self-levelling system proved to work well inside its operating limits of slope, operating gradually to keep the cab horizontal.

From the point of view of the axial accelerations, the horizontal components (X and Y) always resulted in being higher than the vertical one which was never dominant. The activation of the self-levelling system did not show univocal supportive effects, sometimes causing an increase in dominant components, mainly in hillside tests where the presence of high and variable slopes, often exceeding the system's limit, caused its frequent automatic switching on and off. This, combined with soil unevenness, particularly during the in-furrow ploughing, reduced the level of comfort (as the measured values of the accelerations demonstrated). As a consequence, the driver disconnected the system and continued to work in the traditional way. The use of the self-levelling system could be useful in hilly conditions, when the work occurs according to the level lines, or straight down, when the slope is quite constant. An additional, advisable modification should concern the sensitivity of intervention, which should not take into account the punctual soil unevenness (clods) but only the surface general behaviour (slope).

Conversely, during tilling in-plain, in conjunction with the system activation, a certain reduction of the vibration levels was observed, more evident in terms of the resultant acceleration a_v, compared to the traditional condition.

In general, the results show that the tillage is an activity characterized by significant solicitations (vibration and jolts) occurring on all spatial axes: beyond the dominant axial acceleration, the other components often have comparable values. Ignoring them could lead to an important underestimation of T_e. Therefore, the calculation of the exposure time to vibrations during agricultural work should be made based on the vector sum, a_v, which always resulted in being much greater than the axial accelerations a_{wmax}, even multiplied by the factor $k = 1.4$. This would provide a more prudent health risk assessment and a more correct evaluation of the performance of instrumentations and equipment such as the subject of this study. As an example, Table 1 shows that the T_e values based on a_v are, meanly, about 50% of those obtained with a_{wmax}.

The last consideration is about spine posture during the work. If the self-levelling system is working correctly, the Z axis at the seat (perpendicular to the sitting plane) and the actual vertical line coincide. Under conventional working conditions, when the cab is not horizontal, the Z axis at the seat forms a solid angle with the actual vertical line (for instance, a transversal angle of 9.21° during in-furrow ploughing tests in-plain). The driver tends to compensate for this angle by arching the spine, causing unsuitable load distribution on the inter-vertebral disks. In such a situation, it is possible that the X and Y components also contribute to the actual vertical result of vibration. In this case, in order to avoid an underestimation of the risk from exposure to vibration, the X and Y components should contribute to the calculation of the time of daily exposure (T_e). All three axial components a_w are also used in the assessment of the driver's comfort level, which will be the subject of a dedicated study. However, an exhaustive evaluation of these aspects will be possible through a multi-disciplinary approach involving the sectors of agricultural engineering (mechanics) and medicine (occupational medicine, orthopaedics).

Author Contributions: Conceptualization, D.P. and R.F.; methodology, L.F.; software, R.G.; validation, D.P., R.F. and L.F.; formal analysis, L.F.; data curation, L.F. and G.V.; writing—original draft preparation, R.F. and D.P.; writing—review and editing, R.F.; supervision, D.P.; funding acquisition, D.P. All authors have read and agreed to the published version of the manuscript.

Funding: This paper was funded with the contribution of the Italian Ministry of Agricultural, Food, Forestry and Tourism Policies (MiPAAFT) sub-project "Tecnologie digitali integrate per il rafforzamento sostenibile di produzioni e trasformazioni agroalimentari (AgroFiliere)" (AgriDigit program) (DM 36503.7305.2018 of 20/12/2018).

Institutional Review Board Statement: Not applicable.

Informed Consent Statement: Not applicable.

Data Availability Statement: Available data are contained within the article.

Acknowledgments: We thank Cesare Cervellini and Gino Brannetti (CREA-IT) for their precious technical support during the realization of the tillage test.

Conflicts of Interest: The authors declare no conflict of interest.

References

1. Fanigliulo, R.; Del Duca, L.; Fornaciari, L.; Grilli, R.; Tomasone, R.; Pochi, D. Efficiency of an ANC system in the tractor cabin under controlled engine workload. *Noise Control Eng. J.* **2020**, *68*, 339–357. [CrossRef]
2. Park, M.-S.; Fukuda, T.; Kim, T.-G.; Maeda, S. Health Risk Evaluation of Whole-body Vibration by ISO 2631-5 and ISO 2631-1 for Operators of Agricultural Tractors and Recreational Vehicles. *Ind. Health* **2013**, *51*, 364–370. [CrossRef]
3. Nguyen, V.N.; Inaba, S. Effects of tire inflation pressure and tractor velocity on dynamic wheel load and rear axle vibrations. *J. Terramech.* **2011**, *48*, 3–16. [CrossRef]
4. Stojić, B.; Poznanović, N.; Poznić, A. Research and modelling of the tractor tire enveloping behavior. *J. Vib. Control* **2017**, *23*, 290–304. [CrossRef]
5. Pochi, D.; Fanigliulo, R.; Fornaciari, L.; Vassalini, G.; Fedrizzi, M.; Brannetti, G.; Cervellini, C. Levels of vibration transmitted to the operator of the tractor equipped with front axle suspension. *J. Agric. Eng.* **2013**, *44*, *e151*, 752–756. [CrossRef]
6. Kim, J.; Dennerlein, J.; Johnson, P. The effect of a multi-axis suspension on whole body vibration exposures and physical stress in the neck and low back in agricultural tractor applications. *Appl. Ergon.* **2018**, *68*, 80–89. [CrossRef]
7. Deboli, R.; Calvo, A.; Preti, C. Whole-body vibration: Measurement of horizontal and vertical transmissibility of an agricultural tractor seat. *Int. J. Ind. Ergon.* **2017**, *58*, 69–78. [CrossRef]
8. Pochi, D.; Fornaciari, L.; Grilli, R.; Betto, M.; Benigni, S.; Fanigliulo, R. First tests on a prototype device for the active control of whole-body vibrations on agricultural tractors. In *Innovative Biosystems Engineering for Sustainable Agriculture, Forestry and Food Production*; Coppola, A., Di Renzo, G.C., Altieri, G., D'Antonio, P., Eds.; Lecture Notes in Civil Engineering Book, Series; Springer Nature: Cham, Switzerland, 2020; Volume 67, pp. 661–670. [CrossRef]
9. Brunetti, J.; D'Ambrogio, W.; Fregolent, A. Analysis of the vibrations of operators' seats in agricultural machinery using dynamic substructuring. *Appl. Sci.* **2021**, *11*, 4749. [CrossRef]

10. Newell, G.S.; Mansfield, N.J. Evaluation of reaction time performance and subjective workload during whole-body vibration exposure while seated in upright and twisted postures with and without armrests. *Int. J. Ind. Ergon.* **2008**, *38*, 499–508. [CrossRef]
11. Lines, J.; Stiles, M.; Whyte, R. Whole body vibration during tractor driving. *J. Low Freq. Noise Vib. Act. Control* **1995**, *14*, 87–104. [CrossRef]
12. Okunribido, O.O.; Magnusson, M.; Pope, M.H. Low back pain in drivers: The relative role of whole body vibration, posture and manual materials handling. *J. Sound Vib.* **2006**, *298*, 540–555. [CrossRef]
13. Zeng, X.; Kociolek, A.M.; Khan, M.I.; Milosavljevic, S.; Bath, B.; Trask, C. Whole body vibration exposure patterns in Canadian prairie farmers. *Ergonomics* **2017**, *60*, 1064–1073. [CrossRef] [PubMed]
14. Essien, S.K.; Trask, C.; Khan, M.; Boden, C.; Bath, B. Association between whole-body vibration and low-back disorders in farmers: A scoping review. *J. Agromed.* **2018**, *23*, 105–120. [CrossRef]
15. Muzammil, M.; Siddiqui, S.S.; Hasan, F. Physiological effect of vibrations on tractor drivers under variable ploughing conditions. *J. Occup. Health* **2004**, *46*, 403–409. [CrossRef]
16. Cutini, M.; Costa, C.; Bisaglia, C. Development of a simplified method for evaluating agricultural tractor's operator whole body vibration. *J. Terramech.* **2016**, *63*, 23–32. [CrossRef]
17. Scarlett, A.J.; Price, J.S.; Stayner, R.M. Whole-body vibration: Evaluation of emission and exposure levels arising from agricultural tractors. *J. Terramech.* **2007**, *44*, 65–73. [CrossRef]
18. European Community. Directive 2002/44/EC of the European Parliament and of the Council of 25 June 2002 on the minimum health and safety requirements regarding the exposure of workers to the risks arising from physical agents (vibration) (sixteenth individual Directive within the meaning of Article 16 (1) of Directive 89/391/EEC). *Off. J. Eur. Communities* **2002**, *117*, 6–7.
19. Ragni, L.; Vassalini, G.; Fang, X.; Zhang, L. Vibration and noise of small implements for soil tillage. *J. Agric. Eng. Res.* **1999**, *74*, 403–409. [CrossRef]
20. Singh, A.; Samuel, S.; Singh, H.; Kumar, Y.; Prakash, C. Evaluation and analysis of whole-body vibration exposure during soil tillage operation. *Safety* **2021**, *7*, 61. [CrossRef]
21. Pochi, D.; Biocca, M.; Fanigliulo, R.; Gallo, P.; Pulcini, P. Sowing of seed dressed with thiacloprid using a pneumatic drill modified for reducing abrasion dust emissions. *Bull. Insectol.* **2015**, *68*, 273–279.
22. Fanigliulo, R.; Biocca, M.; Pochi, D. Effects of six primary tillage implements on energy inputs and residue cover in central Italy. *J. Agric. Eng.* **2016**, *47*, 177–180. [CrossRef]
23. Fanigliulo, R.; Biocca, M.; Pochi, D. An analysis of eight tillage methods in a silty-clay soil: Proposal for flexible tillage cycles. *INMATEH—Agric. Eng.* **2018**, *56*, 49–58.
24. Fanigliulo, R.; Antonucci, F.; Figorilli, S.; Pochi, D.; Pallottino, F.; Fornaciari, L.; Grilli, R.; Costa, C. Light Drone-Based Application to Assess Soil Tillage Quality Parameters. *Sensors* **2020**, *20*, 728. [CrossRef] [PubMed]
25. *ISO 2631-1:1997*; Mechanical Vibration and Shock—Evaluation of Human Exposure to Whole-Body Vibration—Part 1: General Requirements. International Organisation for Standardisation: Geneva, Switzerland, 1997.
26. Cutini, M.; Brambilla, M.; Bisaglia, C. Whole-Body Vibration in Farming: Background Document for Creating a Simplified Procedure to Determine Agricultural Tractor Vibration Comfort. *Agriculture* **2017**, *7*, 84. [CrossRef]
27. Pochi, D.; Fanigliulo, R.; Pagano, M.; Grilli, R.; Fedrizzi, M.; Fornaciari, L. Dynamic-energetic balance of agricultural tractors: Active systems for the measurement of the power requirements in static tests and under field conditions. *J. Agric. Eng.* **2013**, *44, e84*, 415–420. [CrossRef]
28. Nunes, M.R.; Denardin, J.E.; Pauletto, E.A.; Faganello, A.; Spinelli Pinto, L.F. Effect of soil subsoiling on soil structure and root growth for a clayey soil under no-tillage. *Geoderma* **2015**, *259–260*, 149–155. [CrossRef]
29. Fanigliulo, R.; Pochi, D.; Servadio, P. Conventional and conservation seedbed preparation systems for wheat planting in silty-clay soil. *Sustainability* **2021**, *13*, 6506. [CrossRef]
30. Cutini, M.; Deboli, R.; Calvo, A.; Preti, C.; Inserillo, M.; Bisaglia, C. Spectral analysis of a standard test track profile during passage of an agricultural tractor. *J. Agric. Eng.* **2013**, *44*, e145. [CrossRef]
31. Cutini, M.; Bisaglia, C. Procedure and layout for the development of a fatigue test on an agricultural implement by a four-poster test bench. *J. Agric. Eng.* **2013**, *44*, e81. [CrossRef]
32. Romano, E.; Cutini, M.; Bisaglia, C. Study of posture during plowing operation. Analysis of the pressures to the seat. In Proceedings of the International Conference Rural Health & Ragusa SHWA, Lodi, Italy, 8–11 September 2015.
33. Killen, W.; Eger, T. Whole-Body Vibration: Overview of Standards used to Determine Health Risks. CRE-MDS 2016. pp. 4164–4165. Available online: https://uwaterloo.ca/centre-of-research-expertise-for-the-prevention-of-musculoskeletal-disorders/resources/position-papers/whole-body-vibration-overview-standards-used-determine (accessed on 20 May 2022).
34. Cecchini, M.; Piccioni, F.; Ferri, S.; Coltrinari, G.; Bianchini, L.; Colantoni, A. Preliminary investigation on systems for the preventive diagnosis of faults on agricultural operating machines. *Sensors* **2021**, *21*, 1547. [CrossRef] [PubMed]

AgriEngineering

MDPI

Article

Trampling Analysis of Autonomous Mowers: Implications on Garden Designs

Mino Sportelli [1], Sofia Matilde Luglio [2], Lisa Caturegli [1], Michel Pirchio [1], Simone Magni [1], Marco Volterrani [1], Christian Frasconi [1], Michele Raffaelli [1], Andrea Peruzzi [1], Lorenzo Gagliardi [1], Marco Fontanelli [1,*] and Giuliano Sciusco [1]

[1] Department of Agriculture, Food and Environment, University of Pisa, Via del Borghetto 80, 56124 Pisa, Italy; mino.sportelli@phd.unipi.it (M.S.); lisa.caturegli@agr.unipi.it (L.C.); michel.pirchio@for.unipi.it (M.P.); simone.magni@unipi.it (S.M.); marco.volterrani@unipi.it (M.V.); christian.frasconi@unipi.it (C.F.); michele.raffaelli@unipi.it (M.R.); andrea.peruzzi@unipi.it (A.P.); lorenzo.gagliardi@phd.unipi.it (L.G.); giuliano.sciusco@agr.unipi.it (G.S.)

[2] Department of Agriculture, Food, Environment and Forestry, University of Florence, Piazzale delle Cascine 18, 50144 Florence, Italy; sofiamatilde.luglio@unifi.it

* Correspondence: marco.fontanelli@unipi.it

Abstract: Several trials have been carried out by various authors concerning autonomous mowers, which are battery-powered machines. The effects of these machines on turfgrass quality and energy consumption have been thoroughly investigated. However, there are still some aspects that have not been studied. Among these, random trajectory overlapping is one of the most important. To investigate these aspects, two RTK-GPS devices along with the custom-built software used for previous trials has been upgraded in order to precisely calculate how many times the mower drives over the same spot using random trajectories. This parameter, the number of passages in the same position, was hypothesized to explain the autonomous mower's overlapping and trampling action. The trial has been carried out testing a commercial autonomous mower on three areas with different levels of complexity to assess its performances. The following variables were examined: the percentage of mowed area, the distance travelled, the number of intersections, the number of passages, and the autonomous mower's work efficiency. The average percentage of area mown (average value for the three areas) was 54.64% after one hour and 80.15% after two hours of work. Percentage of area mown was 15% higher for the area with no obstacles after two hours of work. The number of passages was slightly different among the three garden designs. The garden with no obstacles obtained the highest number of passages with an average of 37 passages. The highest working efficiency was obtained in the garden with an intermediate number of obstacles with a value of 0.40 after two hours of work. The estimated energy consumption resulted 0.31 Wh m^{-2} after one hour and 0.42 Wh m^{-2} after two hours of working. These results highlight how the correct settings of cutting time may be crucial to consistently save energy during the long period and may be useful for a complete automation of the maintenance of green areas.

Keywords: turfgrass; RTK; GNSS; precision agriculture; cutting system; path planning; maps; autonomous mower

Citation: Sportelli, M.; Luglio, S.M.; Caturegli, L.; Pirchio, M.; Magni, S.; Volterrani, M.; Frasconi, C.; Raffaelli, M.; Peruzzi, A.; Gagliardi, L.; et al. Trampling Analysis of Autonomous Mowers: Implications on Garden Designs. *AgriEngineering* **2022**, *4*, 592–605. https://doi.org/10.3390/agriengineering4030039

Academic Editors: Marcello Biocca and Roberto Fanigliulo

Received: 27 May 2022
Accepted: 27 June 2022
Published: 1 July 2022

1. Introduction

Precision turfgrass management (PTM) is an extension of precise agriculture and has only recently been taken into account as a way to precisely manage pests, fertilization, salinity, cultivation, and irrigation [1,2]. Managers of complex turfgrass sites currently follow precise management procedures for different areas (i.e., golf course greens, tees, fairways, and roughs), which require specific operations. PTM progress is based on acquiring detailed site information and is moving towards a greater precision and efficiency of input management [3]. Specific field applications are targeted for irrigation, fertilizer dispensation, cultivation, and salinity management procedures [4]. Technological innovations have given a significant contribution in agriculture and management of green areas. The adoption of specific techniques or machines provides for working time and cost reduction [5]. Urban environment improvements are provided by greenspaces through different ecosystems services (i.e., reduction of radiation intensity and heat island effects, carbon sequestration, and so on) [6–8]. Among the variety of operations involved in greenspaces maintenance, turfgrass mowing is the most expensive, energy-consuming, and time-consuming [9,10]. Despite mowing the fact that is mandatory in order to fully exploit turfgrass functionality [11], low-maintenance lawns are required more and more [12]. For instance, autonomous mowers represent a technological innovation that may fill this gap. In Italy, autonomous lawn mowers are appreciated for their time and physical effort saving by allowing people to avoid tedious lawn care. The employment of such machines has shown to improve management sustainability by means of no local pollution, very low noise emissions, and by keeping humans away from allergens [13,14]. Assigning turf maintenance to an autonomous mower may prevent people from developing health issues (i.e., hand-arm vibration syndrome [15]). Moreover, Pirchio et al. [16] highlighted how autonomous mowers gave superior results in terms of energy efficiency compared to gasoline-powered machinery. Autonomous mowers are fully automated, because they can be programmed beforehand to operate every day to obtain optimal turf maintenance and a high percentage of mown area [17,18]. Recently, Global Navigation Satellite System (GNSS) technology has been used to study the performances of autonomous mowers and to try to improve their efficiency. Real-Time Kinematic (RTK) Global Position System (GPS) has shown to provide an accuracy of one centimeter [19,20]. These levels of accuracy provided for localization and mapping derive from multiple sensors data fusion [21]. Thanks to this high positioning accuracy, autonomous vehicles are becoming more and more reliable [22]. Sun et al. [23] used RTK systems to track vehicle movement trajectories and to process many other spatial information, highlighting the potential of this system for these types of operations. RTK-GNSS systems could be applied in many contexts such as residential and urban areas to optimize turfgrass mowing through an improved navigation system. Indeed, Sportelli et al. [24] compared the performances of two autonomous mowers operating with random trajectories and systematic trajectories based on RTK navigation systems. The comparison resulted in a significantly higher performance of the autonomous mower working with systematic trajectories in terms of working efficiency, energy efficiency, and time saving. Remote sensing technologies may be applied to perform a thorough analysis of machines' operative characteristics so as to optimize urban greenspaces management [25]. In this regard, Martelloni et al. [26] studied the random trajectories of an autonomous mower to improve the cutting efficiency in different regular shapes of turfgrass areas, highlighting that a regular design of the working area may help to improve the smooth functioning of robot mowers. Sportelli et al. [27] studied the performances of six autonomous mowers of different sizes in areas with higher or lesser obstacle density. The study highlighted how a larger size of the autonomous mowers and a higher number of obstacles negatively affected the performances of the autonomous mowers in terms of working efficiency. Despite the fact that random trajectories may be a valid solution for areas with a reduced number of obstacles and a regular shape, the frequent overlapping may cause some issues and lead to undesirable drawbacks on turfgrass. Trampling damage, deriving from both direct and indirect mechanical stress, and consequent soil structure damage caused by compaction, are very challenging aspects of turfgrass maintenance [28]. As a result, conducting studies on autonomous mowers' trajectories appears to be highly beneficial in order to optimize turf quality and minimize plant stress. Performing trajectory analysis in terms of compaction resistance may help to define long-term maintenance planning and to assess the specific carrying capacity of the studied turf. The custom-built software allowed to calculate the percentage of area mown, the average forward speed, the average working time, the total distance travelled, and the trajectory intersections. Different trials have been carried out to study the parameters related to the trajectory intersections to evaluate autonomous mowers' overlapping and trampling; however, such parameters are often overestimated and difficult to study among different areas. Thus, an additional function has been provided in this new version of the software to calculate the precise number of passages performed by the autonomous mower in the same spot. The aim of this trial was to test the new version of the custom-built software mentioned in [26] in three different garden designs to have even more precise information concerning autonomous mower operational data. In particular, an analysis of the percentage of area mowed, number of intersections, distance travelled, work efficiency, and number of passages of the autonomous mower was carried out to test the new version of the software. Moreover, the parameters considered allowed to give suggestions for green area design improvements in case of long-term automated maintenance.

2. Materials and Methods

The trial was carried out from April 2021 to May 2021 at the Centre for Research on Turfgrass for Environment and Sports (CeRTES) of the Department of Agriculture, Food, and Environment of the University of Pisa (San Piero a Grado, Pisa, Italy) (43°40′ N, 10°19′ E, 6 m a. s. l.). The trial was performed on a mature cool-season turfgrass stand of tall fescue (*Schedonorus arundinaceus* (Schreb.) Dumort.). The area chosen for the trial was an open area with no buildings or trees close to it, so as not to compromise the data collection. The turfgrass was established on a plot of calcaric fluvisol, which had a coarse silty texture (mixed, thermic, typic xerofluvents with a pH of 7.8 and 2.2% of organic matter). During the trial period, irrigation and fertilization were applied as necessary and no weed or pest control was performed.

2.1. Remote Sensing System and Software

The remote sensing system consisted of two Emlid Reach RTK (Emlid Tech Kft., Budapest, Hungary) devices mounted inside two custom-made cases (the base station and the rover; Figure 1). The main components of this system are extensively described in [26]. The rover was installed on the studied autonomous mower, while the base station was positioned at the edge of the studied area and in the same point during the whole trial. The algorithm used by the RTK precisely calculated the distance between the base station and the rover, called "baseline", with an accuracy of ±7 mm on a horizontal surface.

(a) (b)

Figure 1. Details of the remote sensing system consisting of the two RTK-GPS devices: (**a**) the base station on the side of the studied area; (**b**) the rover installed on the autonomous mower.

Two software packages were used to help extract and display the data. One of the two software packages used for collecting the data was RTKLIB (version 2.4.3), an open-source RTK processing software written by Takasu, (Tokyo, Japan). This software allowed to extract and process the data collected by both devices. The RTKLIB off-line processing generated a .pos file that showed the trajectories carried out by the autonomous mower during the entire work session. The other software package used for data collection was a custom-built software "Robot mower tracking data calculator" (Qprel srl, Pistoia, Italy). This software was used to extract operational data form the spatial data measured during the autonomous mower's work. A brief description of the previous version of the custom-built software is provided hereafter, while a more detailed one can be found in [26]. The custom-built software displayed a two-dimensional map showing the recorded points with an accuracy of 0.05 m and allowed to select a specific area on the map (e.g., the studied area) and the cutting width of the different autonomous mowers. With those settings, the software processed the data in order to retrieve several operational parameters such as: percentage of area mown, number of intersections, distance travelled, total working time, total break time, minimum, maximum, and average forward speed values. In this study, the custom-built software was updated to version 1.8.0.0 and provided some innovation. The first innovation allowed operators to change the cutting width (by steps of 1 cm), which significantly affected the different models of the autonomous mower's paths. Another useful function was the possibility to further process the trajectories' intersections in order to compute the number of times the autonomous mower passed on the same position within the selected area. This last parameter was used to assess the autonomous mower trampling effect on the studied area. Figure 2 shows two of the main functions of the updated version of the software. Figure 2a represents an example of a computation of percentage of area mown. The autonomous mowers trajectories (green lines) are visualized inside the map as the results of recorded coordinates and cutting width of the autonomous mower. The yellow area is directly selected from the map and represents the area of interest. Figure 2b, instead, shows an example of the autonomous mower trampling on the selected area. The software processes the trajectories' intersections to obtain the number of passages in the same position. Those processed data are depicted on a color-scaled map to improve the visual rendering, while a histogram below shows the number of pixels correlated with the number of passages (Figure 2b).

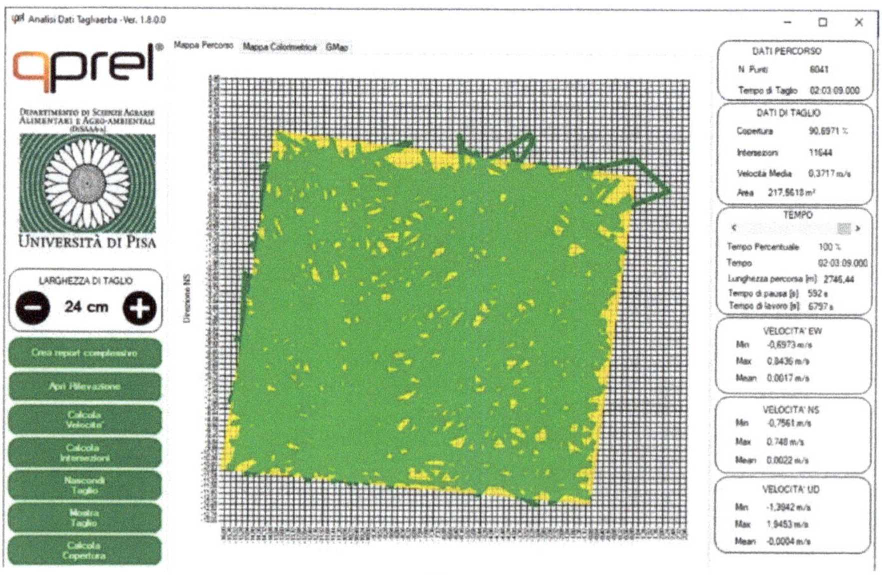

(a)

Figure 2. *Cont.*

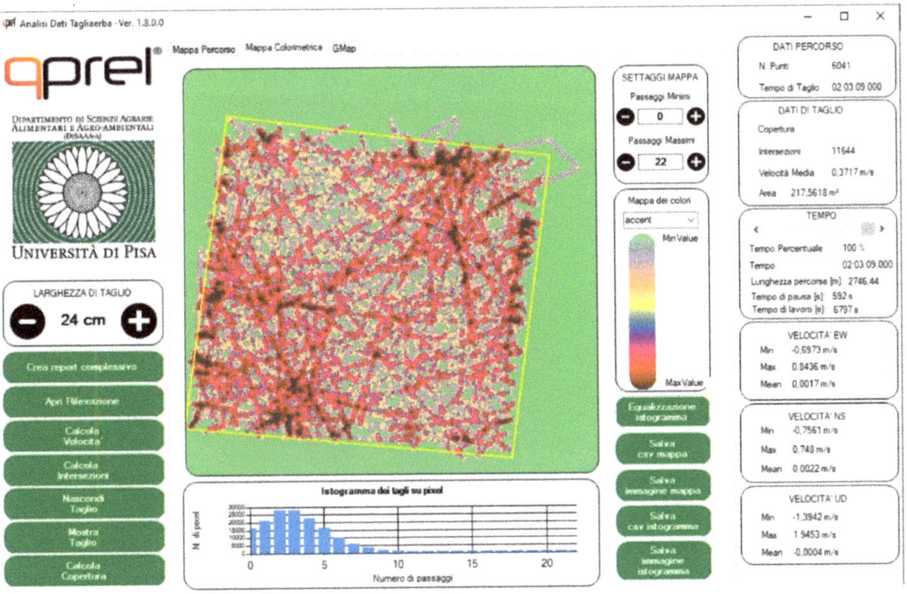

(b)

Figure 2. Software functionality: (**a**) calculation of the percentage of area mown on a square obstacle-free area; (**b**) trampling calculation and visualization on a square obstacle-free area.

2.2. Experimental Field Trials

The experimental plot consisted of a 210 m^2 (15 × 14 m) area and was managed with a Husqvarna Automower® 450X (Husqvarna, Stockholm, Sweden), which moved following straight lines and turned once it intercepted the boundary wire or hit an obstacle (random trajectories). The autonomous mower worked at an average speed of 0.46 m s^{-1} and with a cutting-disk revolving speed of 2300 rpm. The height of cut was set at 3.5 cm, and working width was 24 cm. The area was delimited by a boundary wire that generated an electromagnetic field perceived by the autonomous mower as the edge of the garden. In this trial, in addition to the boundary wire, a guide wire was also used. The guide wire is designed to help the mower to easily find the charging station. The boundary and guide wires were positioned and fixed to the surface with stakes supplied by the manufacturer. The base station remained in the same position during each survey. Obstacles were placed to ensure enough space in each area for the robot to follow the guide wire. In order to avoid interfering with the autonomous mower's movements, it was necessary to leave a clear space of about 1.5 m from the boundary wire. Various garden features (obstacles for the mower) were simulated using wooden poles instead of using the boundary wire to prevent the mower from getting inside such areas. In order to achieve three levels of complexity, some obstacles that simulated common design features were added in the studied area. The obstacles aimed to hinder the autonomous mower so as to better study its performances. The level of complexity changed according to the number and shape of the obstacles. The dimensions of the obstacles were chosen based on product data sheets [29].

The three different garden designs are described hereafter:

- Garden A (Figure 3a) consisted of a 210 m^2 area with no obstacles.
- Garden B (Figure 3b) consisted of a 210 m^2 area with a low level of complexity. To achieve this slight complexity, three obstacles were simulated using wooden poles. The three obstacles were supposed to be a circular bench (diameter 4 m), a 25 m^2 (5 m × 5 m) barbecue area and a rectangular bench (length 2 m × width 0.50 m). In the upper right corner, a circular bench (diameter 0.45 m, thickness 0.50 m and height

0.60 m) and a *Lagerstroemia indica* (L.) tree are placed in the middle (diameter 0.90 m for the roots) and 2 *Delosperma cooperi* (Hook.f.) L. pots (0.70 m^2 and height 0. 60 m).

- Garden C (Figure 3c) consisted of a 210 m^2 area containing all the features of Garden 2 and, in addition, 3 shrubs of *Forsythia* spp. (Vahl) spaced at 1.20 m from each other, a swing of 3 m × 2 m placed close to a slide (length 3.10 m and width 0.40 m) and a lake of approximately 2.30 m^2 of total width.

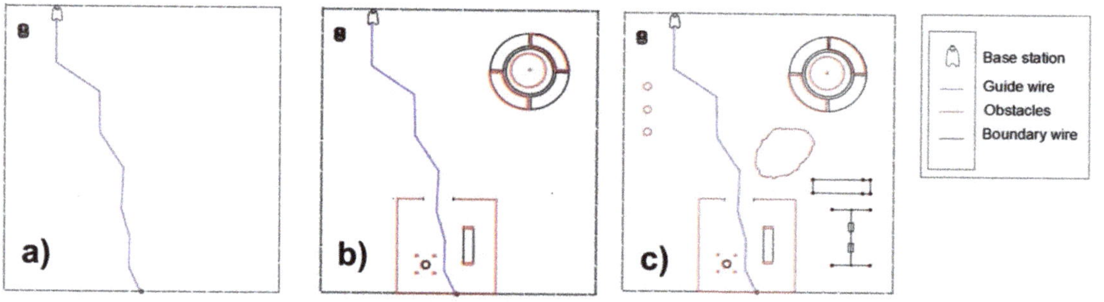

Figure 3. Garden designs scheme studied: (**a**) Garden A, without obstacles; (**b**) Garden B with a circular bench with a *Lagerstroemia indica* L. tree in centre and 2 pots with *Delosperma cooperi* flowers, a barbecue area with a bench, a barbecue, and a shed; (**c**) Garden C with a circular bench with a *Lagerstroemia indica* tree in the centre and 2 pots with *Delosperma cooperi* flowers, a barbecue area with a bench, a barbecue and a shed, 3 plants of *Forsythia* spp., a swing and a slide, 0.40 m wide parts not accessible by the robot, and a small lake.

2.3. Assessements and Statistical Analysis

Each measurement carried out using the remote sensing system lasted two hours. Three runs were carried out for each of the three gardens (for a total of nine runs). Each run was considered a replication for the statistical analysis. Since this research only focused on the operative performance of the autonomous mower and the new functionality of the software, different physical plots were not needed. Before the measurements were performed, the autonomous mower was fully recharged so to obtain an uninterrupted data collection without overestimating the operational data. All the measurements started from the charging station. In this study, the percentage of area mown, the number of passages, the number of intersections, the distance travelled, and the working efficiency were evaluated in function of time. Time was used as a factor with 8 levels (15 min for each time interval). The distance travelled was selected as an operating parameter in order to calculate the working efficiency as indicated by Equation (1):

$$\text{Work Efficiency} = \text{Actual Cut Surface}/\text{Theoretical Cut Surface} \quad (1)$$

where the Actual Cut Surface was obtained by converting the percentage of area mown calculated by the software to its value in m^2. The Theoretical Cut Surface was obtained by multiplying the distance travelled by the autonomous mower with its working width. The percentage of area mown, the number of passages, the number of intersections, and the values of the working efficiency were extracted from the software at intervals of 15 min [30]. The autonomous mowers data were subjected to analysis of variance (ANOVA) using the statistical software R (R Core team, Vienna, Austria). When necessary, data were transformed in order to respect the normality assumption. The values of the percentage of area mown were subjected to angular transformation, while the number of passages, the number of intersections, and the distance travelled were square root transformed. The Shapiro–Wilk test and the Levene's test (package "car") [31] were carried out in order to assess data normality and residual homoscedasticity, respectively. A factorial two-way ANOVA with Garden and Time as independent variables was performed to test the significance ($p < 0.05$) of different garden typologies on the percentage of area mown, on the number of passages, on the number of intersections and on the working efficiency at intervals of 15 min. The ANOVA analysis was followed by post hoc LSD test at the 0.05 probability level provided by the package ("agricolae") [32]. Moreover, an association analysis was carried out in order to evaluate significant positive correlations among the parameters.

3. Results and Discussion

The analysis of variance revealed that the interaction between garden typology and time intervals had a significant effect on the number of intersections, the distance travelled, the number of passages, and the working efficiency (Table 1). The interaction was not statistically significant on the percentage of area mowed.

Table 1. Results of ANOVA testing the effects of garden typology (Garden), time intervals (Time), and their interaction on the percentage of area mowed (Area mowed), the number of intersections (Intersections), the distance travelled (Distance), the number of passages (Passages), and the working efficiency (Efficiency).

Source	Area Mowed (%)	Number of Intersections	Distance Travelled (m)	Number of Passages	Work Efficiency
Garden	***	***	***	***	***
Time	***	***	***	**	NS
Garden × Time	NS	***	**	***	***

***, **, significant at 0.001 and 0.01 probability level, respectively. NS, not significant at 0.05 probability level.

The association analysis revealed that the number of passages was strongly correlated with the number of intersections and the distance travelled producing Pearson's correlation coefficients (r) of 0.921, and 0.856, respectively (Figure 4).

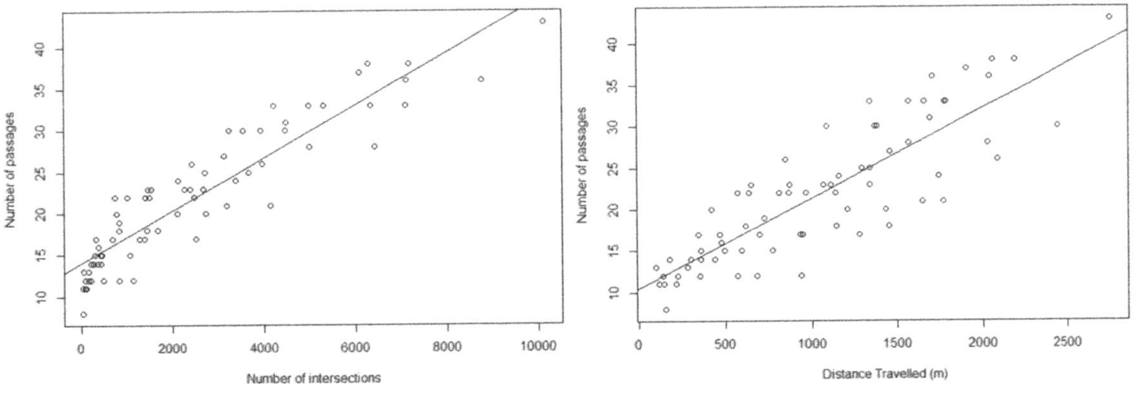

(a) (b)

Figure 4. Positive correlation plots between number of passages and other studied parameters: (**a**) Correlation plot showing positive correlation between number of passages and number of intersections. (**b**) Correlation plot showing positive correlation between number of passages and distance travelled.

Analysis of variance and mean separation tests for the number of intersections and the distance travelled gave similar results to those obtained from the number of passages. Therefore, data concerning the number of intersections and the distance travelled will not be presented and discussed in detail. Figure 5 depicts the percentages of area mowed in function of garden typology and time intervals. Such results were analyzed separately since the interaction between garden typology and time intervals was not significant.

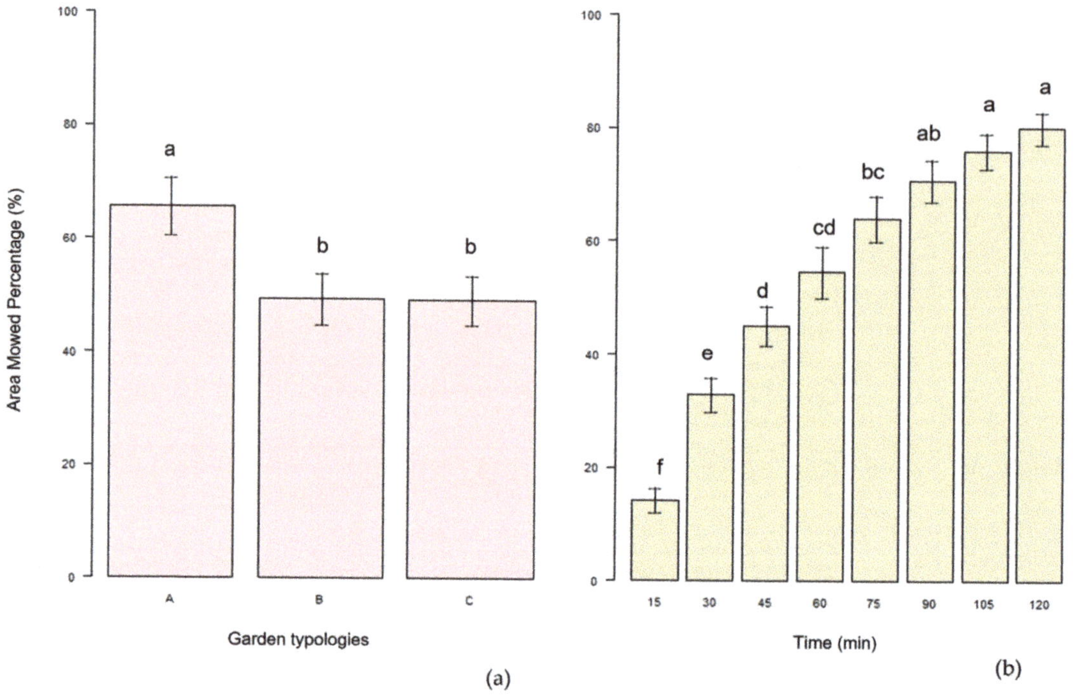

Figure 5. Mean percentage of area mowed for garden typology (**a**) and time intervals. (**b**) Vertical bars represent the standard errors. Different letters on the same plot indicate significant difference at $p < 0.05$ (LSD test).

Mean percentage of area mowed for Garden A was significantly higher compared to the mean percentage of area mowed for the other two gardens with a mean value of 65.67% of the area mowed after two hours of mower activity. No significant differences were observed between the mean percentages of area mowed for the other two garden typologies with an average of 49.34% of the area mowed for garden B and 49.24% for garden C (Figure 5a). Garden A represented an obstacle-free garden, while the other two gardens were arranged with different design features to provide variable garden complexity. Values of the percentage of area mowed after two hours of cutting in obstacle-free gardens with different shapes obtained by [26] were not different, indicating that autonomous mower performances are higher when operating in areas with no obstacles. Such results are in accordance with what has been observed by Sportelli et al. [33]. Authors evaluated the working performances of an autonomous mower working in a vegetable field (with 150 plants established) and compared them with the working performances obtained in a vegetable-free area. The autonomous mower working in the area with the obstacles required an increase of working time of approximately +290% compared to the autonomous mower working in the obstacle-free area. However, in the present trial, the analysis of the percentage of area mowed in function of time intervals revealed that the autonomous mower was able to mow approximately 54.64% of the assigned area after one hour of work and 80.15% of the assigned area after two hours of work (Figure 5b). Figure 6 shows the results concerning the maximum number of passages in function of time intervals and garden typologies (Figure 6).

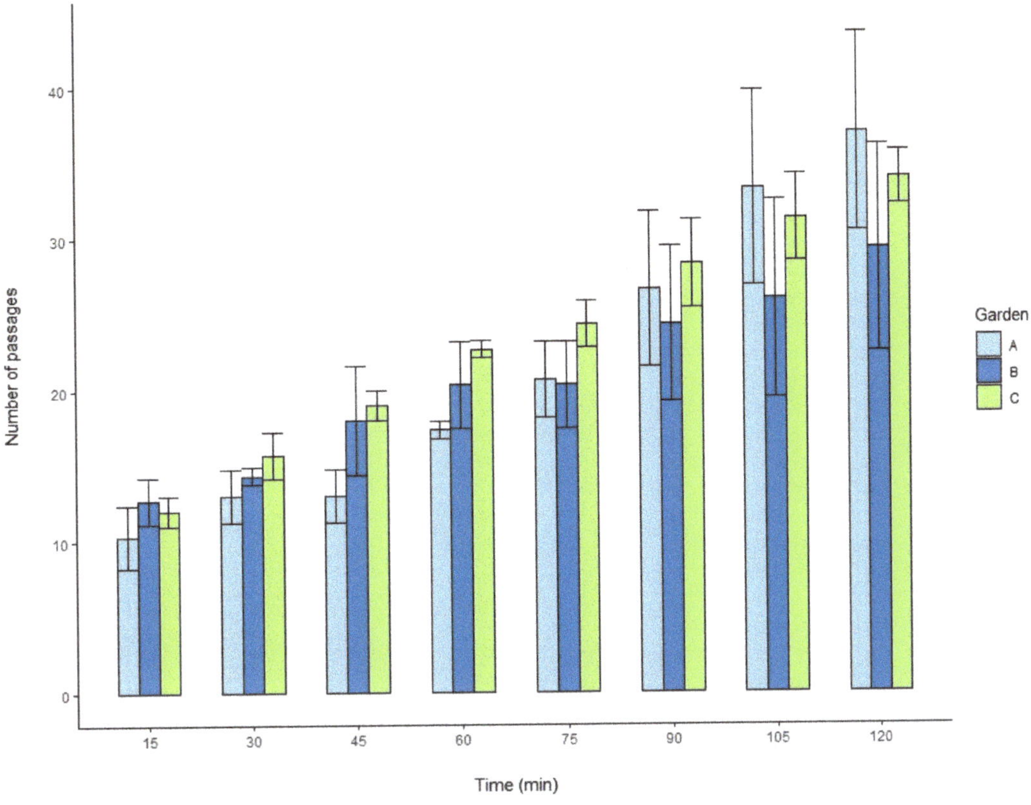

Figure 6. Back-transformed maximum number of passages in function of garden typologies and time intervals. Vertical bars represent the standard errors (A = no obstacles, B = low level of complexity, C = high level of complexity).

After one hour of work, the autonomous mower reached a higher number of passages in Garden C compared to Garden A with an average of 22.67 and 17.33, respectively, while no significant difference was observed compared to the number of passages obtained in Garden B (average value of 20.33). After two hours of work, instead, the higher number of passages resulted from the autonomous mower working in garden A (with an average value of 37). The number of passages measured in Garden B and Garden C after two hours of work did not show significant differences (with an average value of 29.33 and 34, respectively). In general, greater numbers of passages were observed in correspondence with the areas where obstacles were present or with the edges of the gardens (close to the edge, the mower performs more maneuvers since it stops, turns, and departs after changing direction). Figure A1 (in Appendix A) depicts the spatial distributions of the maximum number of passages in the studied areas. Moreover, areas with a high number of passages happen to be not randomly distributed among the different garden types studied. Indeed, a similar trend has been observed in the central areas of garden B and garden C. The reason for this higher number of passages is that, due to its random operating pattern, the mower got stuck in the delimited area and performed a very high number of maneuvers (turning and departing with another direction) before getting out of the area. Moreover, during the mower's first working hour, the mean number of passages observed in Garden A resulted always lower than the number of passages observed in the other two gardens (Figure 6). After the first working hour, the number of passages observed in Garden A started to increase, reaching the highest value after the second working hour. Despite the fact that autonomous mowers are forced to stop and change direction where obstacles are present, overlapping in obstacle-free areas significantly increased over time. The autonomous mower worked more in the areas with no obstacles due to the higher movement capacity, as confirmed by the results of the total distance traveled (data not shown). The distance traveled by the autonomous mower was affected by garden typology. After two hours of work, the distance traveled was significantly lower in Garden B and Garden C compared to Garden A (data not shown). In this regard, as the number of obstacles increases, the area in which the mower can work decreases, leading to shorter paths and an overall lower surface managed. Figure 7 reports the values of working efficiency in function of time intervals and garden typologies.

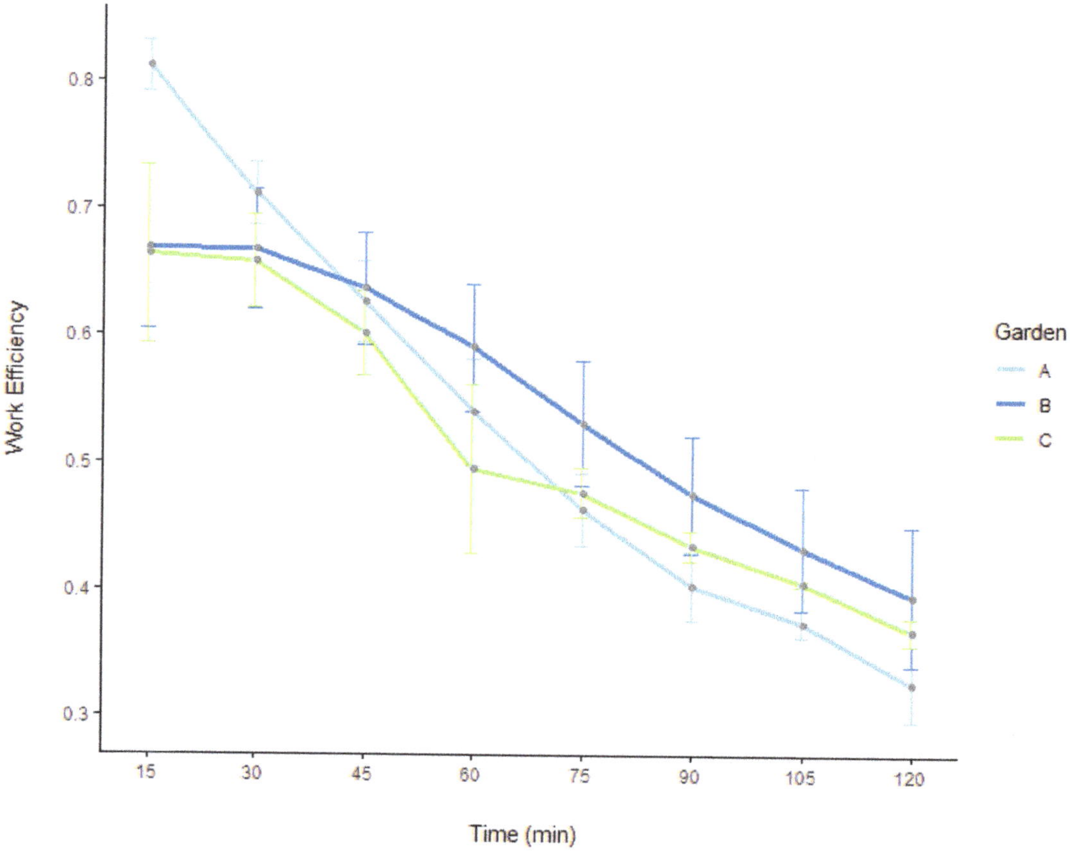

Figure 7. Back-transformed values of working efficiency in function of garden typologies and time intervals. Vertical bars represent the standard errors (A = no obstacles, B = low level of complexity, C = high level of complexity).

The highest working efficiency was observed in Garden A after 15 min of work (0.81). No differences among working efficiency values were observed in the three garden typologies during the first hour of work (values ranging between 0.71 and 0.63). After the first working hour, a significantly higher working efficiency was observed in Garden B compared to Garden C (0.59 and 0.50, respectively). During the second working hour, working efficiency observed in Garden A was always lower compared to the working efficiency observed in Garden B and Garden C (Figure 7). During the second working hour, no significant differences were observed between working efficiency values of the three different garden typologies (values ranging from 0.53 to 0.33). At the end of the second working hour, the autonomous mower's working efficiency was higher for Garden B (0.40) compared to Garden C (0.37) and Garden A (0.33). According to the autonomous mower operator's manual, the studied autonomous mower requires an average power consumption during mowing of 35 W (±20%). The energy consumption per unit area (m^2) was estimated and resulted 0.31 Wh m^{-2} after one hour of working and 0.42 Wh m^{-2} after two hours of working. These values were averaged over the replications and the garden typology. This is in line with the efficiency trends that showed how the overall work performances of autonomous mowers operating with random trajectories decrease during the time. These results highlight how the correct settings of cutting time may be crucial to consistently save energy during the long period, especially for random operating autonomous mowers. Indeed, it is well known that random operating autonomous mowers require more work to manage a given area, due to the frequent overlapping, compared to autonomous mowers' working following systematic trajectories. Working efficiency values observed in Garden A after two hours of work were similar to those obtained by [26], who found that the efficiency of randomly operating autonomous mowers working in an obstacle-free square area was close to 30% after two hours of work. Similarly, Sportelli et al. [24] obtained working efficiency values ranging from 0.32 to 0.35 for a random operating autonomous mower after five hours of work in an obstacle-free rectangular area. In the present trial, Garden B showed the most interesting findings. Indeed, after two hours of continuous work on Garden B, the autonomous mower achieved the highest working efficiency compared to Garden A and Garden C. Garden B represented an intermediate level of complexity between an obstacle-free area and an area with a great number of obstacles. These results highlight that the arrangements of the garden features within an area significantly affected the mower's working efficiency since the autonomous mower operated with the same settings throughout the trial. In this light, Sportelli et al. [25] proposed several design suggestions in order to maximize the efficiency of the autonomous mowers managing green areas and demonstrated how a smart planning can lead to consistent economic savings. In scientific literature, it has been demonstrated that a correct arrangement of garden features may also provide fr environmental benefits. Liu et al. [9] studied the energy consumption and the GHG emission levels deriving from the annual maintenance of an urban greenspace. Such emission levels can be divided into low, average, and high according to the change in plant structure combinations. The emission levels can be useful to help assess the environmental impact of the maintenance of a specific green area. Masoudi et al. [7] studied the effects of urban green spaces composition and configuration on urban heat islands. The authors developed a method to help the correct addressing of greening priorities so as to obtain higher cooling effects. Managing green spaces less intensively has also shown beneficial effects. For example, performing maintenance only on specific areas and not on the total surface may help to reduce mowing operations, energy consumption, and labor costs [34]. This type of management has shown to have beneficial effects on biodiversity and environment. However, in urban areas a more extensive management is usually required, so as to avoid unpleasant aesthetical effects of long vegetation (wild-looking areas). Many attempts have also been made to further maximize the autonomous mowers' efficiency: path planning and systematic trajectories [35]. For instance, Sportelli et al. [24] compared an autonomous mower operating with random trajectories with an autonomous mower operating with systematic trajectories. The trial was carried out on two identical rectangular areas with no obstacles. The working efficiency of the autonomous mower working with systematic trajectories resulted approximately 80%, while the working efficiency of the autonomous mower working with random trajectories ranged from 30% to 35%, approximately. Bosse et al. [36] tested the working efficiency of a large prototype autonomous mower that had a working width of 2 m on an area of 321 m^2. Testing showed that by using a spiral inward operating algorithm (near the perimeter) and a spiral shift operating algorithm (in the inner area), the mower managed to cover 95%

of the area in 15 min. However, this was a very large machine compared to commercial autonomous mowers designed for private gardens such as the one chosen for the present trial, characterized by a working width of 0.24 m. The large working width contributed to reduce the time needed to cover the entire turfgrass area. Another strategy could be to manage green spaces less intensively; for instance, reducing mowing operations, energy consumption, and labor costs by performing maintenance only on specific areas and not on the total surface (i.e., mown paths) [37].

4. Conclusions

The present trial highlighted the potential that the tracking software upgrade might have in terms of green area management. In fact, the upgrade enabled to map the points in the selected green areas and to track the progress of the autonomous mower in three different gardens. The accuracy of the RTK-GPS system and the potential of the software package produced useful data for green area management. Data were gathered both from the percentage of area mowed and from the distance travelled. The results were consistent and in line with predictions after two hours of continuous work. The areas close to the garden features and to the edge of the garden were characterized by a higher number of passages and a higher overlapping fulfilled by the autonomous mower. The results of this trial may be useful to gather data for a complete automation of the maintenance of green areas. However, some limitations of this method should be considered. This kind of garden maintenance is suited for areas far from buildings, vegetation, or other obstacles, so as to ensure that the accuracy of receivers and the data collection is not compromised. Another important issue is the spatial distribution of the obstacles within the area managed by the autonomous mower. When placing an obstacle, the minimum distance from the boundary wire and the guide wire should always be considered so as to allow the machine to move freely. Furthermore, the presence of tight passages should be considered when using an autonomous mower, since these machines have a minimum working width, so as to prevent possible manual operations by the user. The estimated energy consumption highlighted how the correct settings of cutting time may be crucial to consistently save energy during the long period and increase the overall sustainability of green areas management. To conclude, it would be of great interest to carry out a similar study using a robotic lawnmower with a systematic mowing pattern. Probably, using systematic trajectories it may be possible to achieve an easier and more satisfactory level of design and management in terms of relationship between area mown and working time.

Author Contributions: Conceptualization, M.S. and M.F.; methodology, M.F. and M.S.; software, M.S.; validation, M.R., S.M. and A.P.; formal analysis, C.F. and L.G.; investigation, S.M.L. and G.S.; resources, M.V.; data curation, L.C.; writing—original draft preparation, L.C. and S.M.L.; writing—review and editing, M.P. and M.S.; visualization, S.M.L.; supervision, G.S.; project administration, M.V. and M.F. All authors have read and agreed to the published version of the manuscript.

Funding: This research received no external funding.

Institutional Review Board Statement: Not applicable.

Informed Consent Statement: Not applicable.

Data Availability Statement: Not applicable.

Acknowledgments: The authors would like to thank Husqvarna A.B., Sweden, for providing the autonomous mower 450 X and technical support and Qprel srl, Italy, for the technical support. Furthermore, the authors would like to thank Lorenzo Greci and Romano Zurrida from the Department of Agriculture, Food, and Environment (DAFE) of University of Pisa for their technical support.

Conflicts of Interest: The authors declare no conflict of interest.

Appendix A

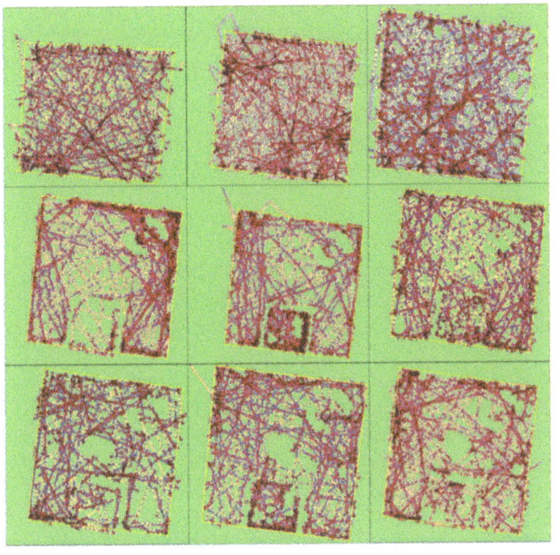

Figure A1. Trampling analysis output from the custom-built software showing colorimetric maps of the three garden designs studied in this trial.

References

1. Stowell, L.; Gelernter, W. Sensing the future. *Golf Course Manag.* **2006**, *74*, 107–110.
2. Bell, G.E.; Xiong, X. The history, role, and potential of optical sensing for practical turf management. In *Handbook of Turfgrass Management and Physiology*; Pessaraki, E., Ed.; CRC Press: Boca Raton, FL, USA, 2008; pp. 641–660.
3. Carrow, R.N.; Krum, J.M.; Flitcroft, I.; Cline, V. Precision turfgrass management: Challenges and field applications for mapping turfgrass soil and stress. *Precis. Agric.* **2010**, *11*, 115–134. [CrossRef]
4. Krum, J.M.; Carrow, R.N.; Karnok, K. Spatial Mapping of Complex Turfgrass Sites: Site-Specific Management Units and Protocols. *Crop Sci.* **2010**, *50*, 301–315. [CrossRef]
5. Magni, S.; Sportelli, M.; Grossi, N.; Volterrani, M.; Minelli, A.; Pirchio, M.; Fontanelli, M.; Frasconi, C.; Gaetani, M.; Martelloni, L.; et al. Autonomous mowing and turf-type bermudagrass as innovations for an environment-friendly floor management of a vineyard in coastal tuscany. *Agriculture* **2020**, *10*, 189. [CrossRef]
6. Tzoulas, K.; Korpela, K.; Venn, S.; Yli-Pelkonen, V.; Kaźmierczak, A.; Niemela, J.; James, P. Promoting ecosystem and human health in urban areas using Green Infrastructure: A literature review. *Landsc. Urban Plan.* **2007**, *81*, 167–178. [CrossRef]
7. Masoudi, M.; Tan, P.Y.; Liew, S.C. Multi-city comparison of the relationships between spatial pattern and cooling effect of urban green spaces in four major Asian cities. *Ecol. Indic.* **2019**, *98*, 200–213. [CrossRef]
8. Keniger, L.E.; Gaston, K.J.; Irvine, K.N.; Fuller, R.A. What are the benefits of interacting with nature? *Int. J. Environ. Res. Public Health* **2013**, *10*, 913–935. [CrossRef]
9. Liu, Y.; Yang, Q.; Duan, L. Adjusting the structure combinations of plant communities in Urban greenspace reduced the maintenance energy consumption and GHG emissions. *J. Environ. Eng. Landsc. Manag.* **2018**, *26*, 261–274. [CrossRef]
10. Jo, H.K.; McPherson, G.E. Carbon Storage and Flux in Urban Residential Greenspace. *J. Environ. Manag.* **1995**, *45*, 109–133. [CrossRef]
11. Beard, J.B.; Green, R.L. The Role of Turfgrasses in Environmental Protection and Their Benefits to Humans. *J. Environ. Qual.* **1994**, *23*, 452–460. [CrossRef]
12. Yue, C.; Wang, J.; Watkins, E.; Bonos, S.A.; Nelson, K.C.; Murphy, J.A.; Meyer, W.A.; Horgan, B.P. Heterogeneous Consumer Preferences for Turfgrass Attributes in the United States and Canada. *Can. J. Agric. Econ.* **2017**, *65*, 347–383. [CrossRef]
13. Ragonese, A.; Marx, J. The applications of sensor technology in the design of the autonomous robotic lawn mower, Paper No. 5094. In Proceedings of the 15th Annual Freshman Engineering Conference, Pittsburgh, PA, USA, 11 April 2015.
14. Hicks, R.W.; Hall, E.L. Survey of robot lawn mowers. In Proceedings of the SPIE 4197, Intelligent Robots and Computer Vision XIX: Algorithms, Techniques, and Active Vision, Boston, MA, USA, 5–8 November 2000.
15. Bernardi, B.; Quendler, E.; Benalia, S.; Mantella, A.; Zimbalatti, G. Occupational risks related to vibrations using a brush cutter for green area management. *Ann. Agric. Environ. Med.* **2018**, *25*, 255–258. [CrossRef]
16. Pirchio, M.; Fontanelli, M.; Frasconi, C.; Martelloni, L.; Raffaelli, M.; Peruzzi, A.; Gaetani, M.; Magni, S.; Caturegli, L.; Volterrani, M.; et al. Autonomous Mower vs. Rotary Mower: Effects on turf quality and weed control in tall fescue lawn. *Agronomy* **2018**, *8*, 15. [CrossRef]

17. Grossi, N.; Fontanelli, M.; Garramone, E.; Peruzzi, A.; Raffaelli, M.; Pirchio, M.; Martelloni, L.; Frasconi, C.; Caturegli, L.; Gaetani, M.; et al. Autonomous mower saves energy and improves quality of tall fescue lawn. *Horttechnology* **2016**, *26*, 825–830. [CrossRef]
18. Magni, S.; Gaetani, M.; Caturegli, L.; Leto, C.; Tuttolomondo, T.; La Bella, S.; Virga, G.; Ntoulas, N.; Volterrani, M. Phenotypic traits and establishment speed of 44 turf bermudagrass accessions. *Acta Agric. Scand. Sect. B Soil Plant Sci.* **2014**, *64*, 722–733. [CrossRef]
19. Ho, V.; Rauf, K.; Passchier, I.; Rijks, F.; Witsenboer, T. Accuracy assessment of RTK GNSS based positioning systems for automated driving. In Proceedings of the 15th Workshop on Positioning, Navigation and Communications, WPNC 2018, Bremen, Germany, 25–26 October 2018.
20. El-Mowafy, A.; Kubo, N. Integrity monitoring of vehicle positioning in urban environment using RTK-GNSS, IMU and speedometer. *Meas. Sci. Technol.* **2017**, *28*, 55102. [CrossRef]
21. Moore, T.; Stouch, D.A. Generalized Extended Kalman Filter Implementation for the Robot Operating System. In *Intelligent Autonomous Systems 13*; Springer: Cham, Switzerland, 2016; pp. 335–348.
22. Xiang, Y.; Li, D.; Su, T.; Zhou, Q.; Brach, C.; Mao, S.S.; Geimer, M. Where Am I? SLAM for Mobile Machines on a Smart Working Site. *Vehicles* **2022**, *4*, 31. [CrossRef]
23. Sun, Q.C.; Xia, J.C.; Foster, J.; Falkmer, T.; Lee, H. Pursuing Precise Vehicle Movement Trajectory in Urban Residential Area Using Multi-GNSS RTK Tracking. *Transp. Res. Procedia* **2017**, *25*, 2356–2372. [CrossRef]
24. Sportelli, M.; Fontanelli, M.; Pirchio, M.; Frasconi, C.; Raffaelli, M.; Caturegli, L.; Magni, S.; Volterrani, M.; Peruzzi, A. Robotic Mowing of Tall Fescue at 90 mm Cutting Height: Random Trajectories vs. Systematic Trajectories. *Agronomy* **2021**, *11*, 2567. [CrossRef]
25. Sportelli, M.; Martelloni, L.; Orlandi, A.; Pirchio, M.; Fontanelli, M.; Frasconi, C.; Raffaelli, M.; Peruzzi, A.; Consorti, S.B.; Vernieri, P. Autonomous mower management systems efficiency improvement: Analysis of greenspace features and planning suggestions. *Agriculture* **2019**, *9*, 115. [CrossRef]
26. Martelloni, L.; Fontanelli, M.; Pieri, S.; Frasconi, C.; Caturegli, L.; Gaetani, M.; Grossi, N.; Magni, S.; Pirchio, M.; Raffaelli, M.; et al. Assessment of the cutting performance of a robot mower using custom built software. *Agronomy* **2019**, *9*, 230. [CrossRef]
27. Sportelli, M.; Pirchio, M.; Fontanelli, M.; Volterrani, M.; Frasconi, C.; Martelloni, L.; Caturegli, L.; Gaetani, M.; Grossi, N.; Magni, S.; et al. Autonomous mowers working in narrow spaces: A possible future application in agriculture? *Agronomy* **2020**, *10*, 553. [CrossRef]
28. Xing, Q.; Qin, J.; Hu, Y.H. Effects of different trampling intensities on three species of warm season turfgrass. *Acta Prataculturae Sin.* **2022**, *31*, 52–61.
29. Design Workshop. *Landscape Architecture Documentation Standards: Principles, Guidelines, and Best Practices*; Wiley: Hoboken, NJ, USA, 2016.
30. Sportelli, M.; Frasconi, C.; Fontanelli, M.; Pirchio, M.; Raffaelli, M.; Magni, S.; Caturegli, L.; Volterrani, M.; Mainardi, M.; Peruzzi, A. Autonomous mowing and complete floor cover for weed control in Vineyards. *Agronomy* **2021**, *11*, 538. [CrossRef]
31. Fox John, W.S. *An R Companion to Applied Regression*; Sage: Thousand Oaks, CA, USA, 2019; Available online: https://socialsciences.mcmaster.ca/jfox/Books/Companion/ (accessed on 26 May 2022).
32. De Mendiburu, F. Agricolae: Statistical Procedures for Agricultural Research. R package version 1.3-5. 2021. Available online: https://CRAN.R-project.org/package=agricolae (accessed on 26 May 2022).
33. Sportelli, M.; Frasconi, C.; Fontanelli, M.; Pirchio, M.; Gagliardi, L.; Raffaelli, M.; Peruzzi, A.; Antichi, D. Innovative Living Mulch Management Strategies for Organic Conservation Field Vegetables: Evaluation of Continuous Mowing, Flaming, and Tillage Performances. *Agronomy* **2022**, *12*, 622. [CrossRef]
34. Norton, B.A.; Bending, G.D.; Clark, R.; Corstanje, R.; Dunnet, N.; Evans, K.L.; Grafius, D.R.; Gravestock, E.; Grice, S.M.; Harris, J.M.; et al. Urban meadows as an alternative to short mown grassland: Effects of composition and height on biodiversity. *Ecol. Appl.* **2019**, *29*, e01946. [CrossRef]
35. Wang, J.; Chen, J.; Cheng, S.; Xie, Y. Double heuristic optimization based on hierarchical partitioning for coverage path planning of robot mowers. In Proceedings of the 12th International Conference on Computational Intelligence and Security (CIS), Wuxi, China, 16–19 December 2016.
36. Bosse, M.; Nourani-Vatani, N.; Roberts, J. Coverage Algorithms for an Under-actuated Car-Like Vehicle in an Uncertain Environment. In Proceedings of the IEEE International Conference on Robotics and Automation, Rome, Italy, 10–14 April 2007.
37. Hoyle, H.; Jorgensen, A.; Warren, P.; Dunnett, N.; Evans, K. Not in their front yard. The opportunities and challenges of introducing perennial urban meadows: A local authority stakeholder perspective. *Urban For. Urban Green.* **2017**, *25*, 139–149. [CrossRef]

www.ingramcontent.com/pod-product-compliance
Lightning Source LLC
Chambersburg PA
CBHW061558210926
51850CB00009B/2138
9783725805488

MDPI
St. Alban-Anlage 66
4052 Basel
Switzerland
www.mdpi.com

AgriEngineering Editorial Office
E-mail: agriengineering@mdpi.com
www.mdpi.com/journal/agriengineering